WINNIPEG

WN

D0594092

"Venkat's book is super helpful to all of us working to drive change in an increasingly digital world."
MIKE WRIGHT Global CIO, McKinsey & Company

"For any CXO executive, this book is a must-read. Professor Venkatraman brilliantly describes a practical step-by-step guide on how to transform a legacy company into a digital enterprise. An essential survival guide for today's evolving digital economy."
ALEJANDRO MARTINEZ Senior VP and CIO, Quintiles

"A refreshing read from a scholar who has been at the forefront of strategy and digital technology for over two decades."
RANJAY GULATI Professor, Harvard Business School, author, *Reorganize for Resilience*

"No firm, no industry and no market is immune from the transformative forces that disrupt the practices of the past, and we may have no better guide through these forces than Professor Venkatraman. No enterprise leader should be without this useful guide."
BENN KONSYNSKI PhD, Professor, Emory University

"This book is definitely the best I have read dealing with the digital challenges and how to address them."
JO GUEGAN Strategic IT Advisor, Former CTO and EVP of Canal+ Group, Former SVP, Capgemini Consulting

"Venkat Venkatraman enthralls with a sense of urgency and offers a practical and composed approach to assess threats and devise winning and competitive strategies."
DON BULMER VP, Gartner; Former VP, Shell, SAP

"This book contains invaluable insights and should be required reading for executives to step beyond their industry boundaries. I can't wait to apply these ideas in practice."
GEORGES EDOURAD DIAS Co-founder and CSO, Quantstreams, Former CDO, L'Oreal Paris

"*The Digital Matrix* captures the different challenges and opportunities of pursuing a disruptive digital strategy for a traditional business. All leaders must be aware of how digital can transform their organization if they want to thrive in this new era."
TIM THERIAULT Former Global CIO, Walgreens Boots Alliance, Former President of Corporate and Institutional Services, Northern Trust

"*The Digital Matrix* provides a framework for participation, learning, and building relationships across all phases of transformation. No longer just in the realm of innovation, the practical application of this framework needs to be an active and present part of corporate planning processes."
JIM CIRIELLO AVP, IT Planning & Innovation, Merck

"*The Digital Matrix* offers compelling insight into why traditional organizations should embrace digitization and stay vigilant to the signals at the periphery."
MAHESH AMALEAN Chair, Co-founder, MAS Holdings

"To Venkatraman, Facebook and Tesla are not traditional companies bound by product boundaries or SIC codes; they adapt digital tools to solve problems. This book gives you a strategy roadmap to stand alongside such problem solvers, to anticipate their next move, to compete against them. Without it, you risk being wiped out by them."
BHASKAR CHAKRAVORTI Senior Associate Dean, International Business & Finance, The Fletcher School at Tufts University, author, *The Slow Pace of Fast Change*

"*The Digital Matrix* is not just another descriptive or speculative account about future technologies and their predicted impacts—it is the definitive guide to becoming a proactive player in the new digitally meditated economy. This book will survive the passage of time."
BEN M. BENSAOU Professor, INSEAD, Fontainebleau

"*The Digital Matrix* should be read as a clarion call to executives and board members. For the new wave of leaders, this book will be a go-to guide."
RICK CHAVEZ Partner, Digital Practice, Oliver Wyman

"No silver bullets, no killer apps. Instead, Venkatraman provides a brilliant exposition on the perfect storm of digital technologies that will severely test the leaders of every organization, and a framework for analysis and action to help us survive and thrive in the coming decade."
BRINLEY N. PLATTS Chairman, CIO Development

"Venkatraman is a brilliant thought leader in the area of digital business strategy. This book will definitely turbocharge your digital future!"
OMAR EL SAWY Chaired Professor, USC
Marshall School of Business

"This is an important and timely book. With deep familiarity, clear examples and nuance, Professor Venkatraman offers a much-needed, sophisticated roadmap for incumbent firms to leverage digitization and thereby prosper in our new competitive era."
MEL HORWITCH University Professor and Former
Dean, CEU Business Administration

"*The Digital Matrix* is a wake-up call for any business intersecting with the digital world. Health care leaders need to heed this call and consider how they must lead in the increasingly digital world."
CHRIS NEWELL Psy.D, Director, Learning and
Development, Boston Children's Hospital

"Understanding these insights is essential if we are to clearly see the challenges and opportunities created by the digital revolution. Venkat Venkatraman challenges our conventional thinking and encourages us to become transformational leaders in our own fields of endeavor."
MIKE LAWSON Professor Emeritus, Boston
University Questrom School of Business

"Venkatraman perfectly blends academic and consultant into one clear roadmap for leading change, enabling leaders to act immediately after putting the book down."
STEVE NEWMAN Former Director, Executive Programs at Ericsson

"*The Digital Matrix* is a brilliant window into digital strategy, with practical insights that blend academic theory and the practice of management in a way that only Venkatraman can deliver."
JOHN C. HENDERSON Professor Emeritus, Boston
University Questrom School of Business

"This book is a powerful guide to the most important change to management of the last century—the digital transformation of every organization. It offers a practical vision of what it will take not only to adapt to the networked society, but to lead it. An enlightening book and a terrific read!"
RICHARD LEIDER bestselling author, *The Power of Purpose*

THE DIGITAL MATRIX

New Rules for
Business Transformation
Through Technology

VENKAT VENKATRAMAN

A LifeTree Media Book

Copyright © 2017 by Venkat Venkatraman

17 18 19 20 21   5 4 3 2 1

All rights reserved. No part of this book may be reproduced,
stored in a retrieval system or transmitted, in any form
or by any means, without the prior written consent of the
author and/or publisher or a license from The Canadian
Copyright Licensing Agency (Access Copyright). For a
copyright license, visit www.accesscopyright.ca or call
toll free to 1-800-893-5777.

Published by
LifeTree Media Ltd.
www.lifetreemedia.com

Distributed by
Greystone Books Ltd.
www.greystonebooks.com

Cataloguing data available from Library and Archives Canada
ISBN 978-1-928055-20-4 (hardcover)
ISBN 978-1-928055-21-1 (epub)
ISBN 978-1-928055-22-8 (pdf)

Editing by Lucy Kenward
Cover design by David Gee
Interior design and illustrations by
    Setareh Ashrafologhalai
Printed and bound in Canada by Friesens
Distributed in the US by Publishers Group West

To my parents, who have always believed that
I would do my best in whatever I chose to do.
To my late in-laws, who treated me as
the son they didn't have.

To my wife, Meera, who has always supported me
in everything I have pursued and done.

To my daughters, Tara and Uma, whose lives
will be more digital than we can ever imagine today.

CONTENTS

PREFACE

○─────────○

I AM A STUDENT of business. For more than three decades, I have been studying organizations to understand how they work, what drives their success, and where their vulnerabilities are. I analyze companies in terms of what they are capable of doing and guide them to achieve new levels of efficiency and pursue new trajectories of growth. In that sense, I am not unique. Every business school professor, irrespective of functional focus (e.g., marketing, operations, strategy, accounting) or disciplinary orientation (e.g., economics, psychology, sociology, computer science) is interested in the questions of short-term efficiency and long-term effectiveness. Where I differ from others is in my interest in how management as we know it—as we teach, research, and practice it—is affected by digital technologies of different types and in different ways.

When I began my academic career at the MIT Sloan School of Management, the business school of the Massachusetts Institute of Technology, in 1985, I was fortunate to be part of a collaborative research initiative called Management in the 1990s, led by Professor Michael Scott Morton and sponsored by many leading commercial organizations and government institutions. Our mandate was to answer the following question: "How can businesses transform themselves by taking advantage of the power of information technology, and what could it mean for the discipline of management as we know it?" At that time of IBM personal computers, Xerox Star workstations, and Digital Equipment mini computers, we talked about "information technology" (IT) or "information systems" (IS). IT-driven innovations included automatic teller machines (ATMs), emails, and industry-specific protocols and communications that linked firms via dial-up modems! And Nicholas Negroponte, the founder of *Wired* magazine, had just published his prescient bestseller *Being Digital* about digital technologies and their possible futures.[1]

As I looked into how companies such as General Motors (now GM), American Express, British Petroleum (today's BP), the Internal Revenue Service (IRS), and the US Army were using some of those early innovations, I developed a framework that outlined five levels of IT-enabled transformation. My thesis at that time, which I published in the MIT *Sloan Management Review* in 1994, and which still stands as one of the journal's bestselling reprints, was simple and straightforward: to realize benefits from IT applications requires corresponding changes in how firms are organized and how they interconnect with others in extended business networks. Moreover, whereas mid-level managers can deal with localized ways to take advantage of information systems, senior managers must understand how IT can change the very essence of their company's strategy, namely its scope of business operations, its core capabilities (which differentiate one firm from another), and its make-buy-partner decisions.

The prevailing wisdom at that time was that IT strategy should respond to business strategy, that IT strategy and organization must be set up to support the chosen business strategy. Based on our research, however, my colleague John C. Henderson and I developed a new business logic. We called upon managers to think systematically about the conditions in which IT strategy *supported* business strategy and those in which IT strategy *shaped* the business strategy itself. This logic, the supporting framework, and the management recommendations formed the basis of the Strategic Alignment Model, which was first published in the IBM *Systems Journal* in 1993 and again in 1999, when it was declared one of the key ideas, or "turning points," for thinking about IT from the 1960s until the end of the twentieth century.

First at MIT and now in the Questrom School of Business at Boston University, my research, teaching, and consulting work have continued to trace the intersection of business and IT strategies. I have looked at how firms interconnect across boundaries with information technology in settings such as the airline, insurance, automotive, and retailing sectors. With John Henderson, I have developed ideas about partnerships—especially to construe a corporation as a portfolio of capabilities assembled through networks of relationships. And I have also researched how these "ecosystems" begin, grow, and evolve in two areas—software and video games—and studied the emergence of social, mobile, and media webs. I see companies as interconnected, and the successful ones know when to be linked at the core to emerging ecosystems and when to pull back. I have refined and tested these ideas with L'Oréal, Canal+ in Paris, Nielsen, IBM (and its clients), Microsoft, British Telecom (BT), Swire, BP, Statoil, Sears Canada, Lucent, GM, Merck, the US Army, FedEx, Visa Europe, Ericsson, and many others, and come to understand how companies co-create their offerings and capture value working in these dynamic business networks.

Today, digital technology is everywhere. Google and Facebook are global brands. Typing skills are being replaced by voice commands; printed road atlases, encyclopedias, and dictionaries have been supplanted by online databases; photo albums and CDs have been rendered obsolete by cloud storage. Soon, self-driving cars will transport us and drones will make our deliveries. Robots are moving from factories into our homes, our offices, and our hospitals. Although we have seen a significant increase in the pervasiveness and the power of digital technology, many businesses have not kept pace with the change. Three decades on, the core question guiding my work remains the same: "How can businesses transform themselves by taking advantage of digital technologies?" What makes some companies embrace emerging digital technologies while others find themselves like a deer staring at the headlights? Why do most executives intuitively appreciate the power and promise of digital technologies but still shy away from immersing deeply enough to understand how best to use them to advantage? Why do business managers in traditional industries say that the future will be different but continue to rely on their established playbooks?

My goal in writing this book is to help guide managers to take digital technologies more seriously. First, although there have been many books about the success of new digital "shining stars," such as Google, Facebook, eBay, Amazon, and tech entrepreneurs, such as PayPal,[2] and books about important trends, such as the social web, mobile web, big data, sharing economy, future of work, etc., there has been little serious and systematic attention paid to what incumbent businesses in traditional industries can (and should) do and how they might interact with these digital stars and technological trends. Second, I believe we are at a tipping point when we will witness dramatic changes not only in the raw power of digital technologies in new domains such as cloud computing, robotics, 3D printing, machine learning, blockchain algorithms, and the like but also, and more importantly, in how they apply to the core activities of how humans live, work, learn, play, innovate, transact, and govern. I believe that digitization will pervade the very fabric of our society, and during this period of transformation, we will see winners and losers defined by their ability to master these technologies.

I do not claim to be a futurist—and I cannot predict what the business environment will look like in 2025—but I believe that the impacts on every industry will be stronger and deeper than simply introducing mobile apps and social interactions or hiring data scientists to sift through and analyze volumes of data. To this end, I have written this book for executives of legacy businesses in traditional industries, such as manufacturing, automotive, mining, oil and gas and energy, transportation and logistics, consumer products and retailing, travel and hospitality, pharmaceuticals and health care, fashion and apparel, and so on. I cannot think of a single setting in which digitization doesn't hold promise today or is unlikely to matter in the future. I provide a framework to help you understand digital business strategy; to make you think and question your own current business logic and practices, assumptions, and assertions; and to guide you in assessing and making changes. It doesn't matter at what level or function you are in the company—or whether you are in a large, small, or a medium-sized company—you could play a critical role in the digital transformation of your company. It really doesn't matter much if you are located in North America or Europe or India or China or Korea or Sub-Saharan Africa or South America. Digitization is a global trend and there are no safe sanctuaries in which to shelter outmoded business models.

Although I may use larger companies as my examples, this is mostly so that you can recognize them readily. Remember that the future is less about the size of individual companies but more about how organizations of different sizes work together to take advantage of the power of digital technologies to deliver value to customers. I hope this book inspires you to delve deeper into understanding what the future holds, what the next decade may bring, what the digital decade may unleash in terms of exciting innovations and agonizing heartaches for companies that miss its opportunities. Even in 2025, we will probably still be asking: "How can businesses transform themselves by taking advantage of the power of digital technologies, and what could it mean for the discipline of management as we know it?" The only difference is that the set of technologies will be new and the management principles and practice will have changed. My hope is that you will be one of the leaders who has successfully guided your business through digital transformation and that your journey will form the basis for how we think about the future of management in 2025.

INTRODUCTION

YOUR BUSINESS... HANDICAPPED BY AN OLD RULEBOOK

BUSINESSPEOPLE LOVE TO talk about the *Fortune* 500. We love to debate what this annual published list of US companies with the highest revenues implies about the economy as a whole. And more than anything, we love to tease out trends and try to draw conclusions from them. So when I say that sixty-one companies on the magazine's original list from 1955, about 12 percent, were still on the list in 2015,[1] what does it mean? Or that fifty years ago, a company on the *Fortune* 500 list could expect to last sixty-plus years, but now, that life expectancy is just fifteen to twenty years.[2] It suggests to me the caveat made famous by the fine print associated with mutual funds: "Past performance is no indicator of future returns." In other words, you cannot rely on your existing business strategies—no matter how successful—to carry you into the future. If you are still working with industrial-age ideas, you are, or very soon will be, out of step with the reality of the digital economy. And that's precisely why you need to read this book.

The evidence is unmistakable. Since the mid-1990s, when it began as a search-related startup company imagined by two geeky computer scientists, Google has become a global technology brand worth more than half a trillion dollars. Yes, that's trillion with a t. In two decades, Amazon has gone from being an online bookstore to a global e-commerce retailer that has overtaken Walmart in sales, as well as a leader in cloud computing services, media, and artificial intelligence. And Facebook, which began as an online social networking site available only to Harvard University students, has become a global social network worth more than $300 billion. And during the past decade, it has been facilitating conversations between brands and consumers and between electoral candidates and voters. These three relatively young companies—the oldest of the trio is just twenty-one years old—were capitalized at more than $1 trillion by mid-2016. They have succeeded by taking advantage of impressive developments in digital technology to craft powerful business models and to become large powerful digital giants. Where do these companies rank on the *Fortune* 500 list? Facebook has been on the list for just four years and has jumped from #482 in 2013 to #157 in 2016. Google (now Alphabet) leapt onto the list at #353 in 2006 and stands at #36 in 2016. Amazon made the list for the first time in 2002 at #492 and has climbed steadily to #18 in 2016. Apple's positioned at #3, Microsoft at #25, and IBM at #31, all based on revenues. And one day in early August 2016, the top five market capitalized companies on the Dow Jones index were: Apple, Alphabet, Microsoft, Amazon, and Facebook.

What lessons do these companies offer? That digital technology is critical to every industry and every company, including yours. Google, Amazon, Apple, and Facebook were not—and are not—traditional information technology (IT) companies. Digital technologies help these companies become powerful global brands—according to Interbrand's list of the world's best brands for 2015, Apple is #1, Google is #2, Amazon is #10, and Facebook is #23[3]—which allows them to reach more people around the world more quickly than ever before.

So let's debunk two common misconceptions. First, if I asked, "Is your business digital?" you, like many executives, would probably quickly say no, that your business is ultimately about health care or transportation or hospitality or fashion. And then if I asked, "If your business is not digital, which business is?" you, like many others, would probably say "high-tech companies" such as Google, Amazon, and Facebook or larger global players such as Apple, Alibaba, and AT&T, or even startups such as Airbnb, Uber, Twitter, and Fitbit. You might believe digital is increasingly *relevant* for you but that it is not *critical*. Let me say it again, digitization is critical for every business, in every industry, everywhere *now*. This technology is pervasive, yet most executives are the least knowledgeable about and the most unprepared for the transformations it can and will bring about.

DIGITAL IS EATING THE WORLD

We are in the early stages of this digital transformation and cannot know what the future looks like. Technical people love to talk about the latest and greatest technologies on the horizon. You have doubtless been inundated with presentations about mobile, social, cloud, and cognitive technologies. To that list, we can add 3D printing, robotics, blockchain, artificial and augmented intelligence, drones, nanotechnology, and virtual reality. Some are here now, others are a bit further afield, and still others will be added next year and in the years after that. We can say with certainty: "Digital is eating the physical, industrial world."[4]

The reordering of the business leaderboard—represented by the *Fortune* 500 or other lists—will increasingly be influenced by digital technologies. But this book is not about a technology push. If I have learned one lesson in my research, teaching, and consulting, it is this: the benefits from new digital technologies accrue only when companies change their organizational architecture—their structure, process, decision rights, interfirm relationships, resource allocation logic, incentives, and rewards.[5] Simply overlaying technology, however powerful, on existing organizational architectures does not work. By going digital, I mean embracing the new business infrastructure that's at the intersection of, as I like to say, "powerful computing, pervasive connectivity, and potent cloud." We have seen computing power grow exponentially and at lower and lower cost (Moore's Law[6]), we have seen the value of networks grow as they increase in size (Metcalfe's Law[7]), and we have seen more data being transmitted with greater reliability and at reduced cost using cloud computing technology (Gilder's Law[8]). These three forces acting together have created the new business infrastructure, and this is the new reality to which your firm must adapt. So your task is not just to automate traditional manufacturing and administrative processes but to use digital technologies to learn about and solve customers' key problems and to work with others within and across traditional company and industry boundaries.

Hans Vestberg, the chief executive officer of the global telecom company Ericsson until 2016, is fond of saying, "Anything that can benefit from a connection will have one in the Networked Society." Restating it slightly, my view is that any product, process, service, or business infrastructure that could benefit from being digitized will be digital in the near future. As a business leader, you do not have to become an expert in the ins and outs of each technology, but you must develop the necessary acumen to know how applying different technologies could challenge your business model—altering the sources of revenue and shifting the sources of profits. You need to shift your attention from thinking about how digital technologies *support* your current business to examining how they could also *shape* your future strategy and business models. To do this, you need to examine broadly and evaluate prudently how you could take advantage of technology developments and digital partnerships to craft innovative new ways to create and capture business value.

WHY IS DIGITIZATION NOT YET ON THE AGENDA?

To understand why you should focus on digitization now, we must start with a broader but very central question: "Why do *successful* companies die?" And I believe the answers serve up the reasons that limit companies from recognizing the opportunities and challenges posed by digitization. I call these reasons "success traps" because they are all strategies that have allowed individual companies to compete and thrive in industrial-age economies with the traditional business models that you know well.

The competency trap

Every company develops its core competencies, first by acquiring a set of physical assets and then by investing in people with the knowledge and skills to create its products and structure the organization with appropriate governance. Those core competencies get progressively more refined and enhanced with the entire organization's strategies, structures, and systems aligned around them. They define how a company earns revenue and profits, they differentiate one company from another, and they are the reason customers will pay a premium price for a particular product or service. Over time, these competencies become difficult for competitors to imitate, and these competencies are then considered core to the company's future performance. The result is that business leaders stay with the current business model because it works. But there's little opportunity to question when or whether these competencies might run their course—that is, when or whether there may be a point at which customers no longer value those competencies and profits will fall.

Take BlackBerry and Nokia as examples. In 2007, BlackBerry thought it had the perfect smartphone for corporate business executives. It had an important messenger app for email, software optimized for limited bandwidth, and a battery optimized for long usage. Around the same time, Nokia thought that its feature phones with short message service (sms, or text messaging) and global reach could guarantee its future success. They were both leaders in their respective domains. Then Apple introduced the iPhone, and BlackBerry and Nokia's phones were rendered obsolete. Both companies were simply trapped in their historical competencies. More companies will be trapped similarly, especially during periods of digital transformation.

This book shows you ways to think beyond your current competencies and position yourself for relevance in the digital future.

The ecosystem trap

Every company develops networks of relationships with supply chain partners, technology vendors, key marketing partners, research and development and innovation companies, and many others. These relationships are built over time through trust and complex negotiations, and many companies develop specific organizational processes to maximize these relationships and reinforce the core competencies. The result is that companies stay with the same suppliers and partners because they work. But there's little time to question whether other relationships might bring better value to the company, especially when they could conflict or challenge existing arrangements. In digital-era companies, however, these relationships are key: you need to rely on your traditional competitors as well as technological startups and large digital companies. These networks of companies, called ecosystems, both cooperate and compete with each other in whole new ways from how traditional businesses are used to collaborating.

Take Microsoft, for example. Bill Gates masterfully orchestrated Microsoft's dominance in the personal computing ecosystems in the 1990s. The company developed strong relationships with hardware manufacturers such as Samsung, HTC, Sony, and Toshiba to put a "computer on every desk and in every home." The company was focused on fine-tuning its ecosystem to keep up with new versions of its software running on ever-faster computers powered by more and more powerful Intel chips. Then Motorola and Nokia introduced mobile phones, and Apple came out with its smartphone, essentially a handheld computer that worked nearly everywhere. Even though Steve Ballmer, who succeeded Gates, had relationships with companies that were part of the emerging mobile ecosystem, he did not recognize the value of a mobile operating system ecosystem. He and Microsoft were simply caught up in preserving their existing relationships, and Microsoft missed the boat on mobile.

This book shows you ways to structure and nurture ecosystems that cut across industry boundaries, especially as your industry's traditional disciplines—mining, agriculture, health care, transportation, and others—collide with digital technologies.

The talent trap

Every company strives to acquire, nurture, and manage the best human talent and capture value from its expertise. It's a complex business, especially striking the balance between deep expertise in specific domains to be excellent and create differentiation and adapting to changing contexts. The result is that companies hire staff for existing roles instead of identifying the profile of talent that might be needed in the digital future. And there's little attention to nurturing digital business literacy and bringing in staff with the vision and technical skills to guide and implement new processes and structures that enable digital transformation.

Take Motorola. How did this company, which invented the mobile phone, miss the smartphone innovation even as it worked with Apple to introduce iTunes to phones? Ed Zander, who ran Motorola during the period of transformation, remarked that he saw the smartphone coming, but "Motorola didn't have the DNA or the people to understand the software involved."[9] He was simply caught with an inflexible talent base: his hardware engineers could not become software engineers overnight. According to a survey conducted by the global management consulting firm McKinsey & Company in 2015, the "most common hurdle to meeting digital priorities, executives say, is insufficient talent or leadership."[10]

This book shows you how human talent can work with powerful machines to create new ways of organizing that reflect the digital age where data and analytics at all levels become central and crucial.

The metrics trap

Successful organizations are particularly driven by metrics, most often quantitative measures of efficiency, quality, cost, and profit margins. Even performance is measured against a set of targets, and individual managers, and teams and organizational units are driven to meet specific targets. I have found that most such metrics, all across a company, focus on short-term performance such as market share, sales per unit, or customer profitability. There's nothing wrong with these, except in the absence of longer-term thinking about digital transformation and innovation, they reinforce a near-term focus and incremental changes in how and where scarce resources are allocated. When market share, for example, is the paramount goal, mergers and acquisitions favor familiar companies in conventionally defined industry boundaries rather than digital companies that might bring much-needed newer capabilities. If the automotive companies focused less on "the number of vehicles sold" and more on "the share of people-miles traveled," how might they design their business?

Edwards Deming, one of the foremost management scientists of the twentieth century, remarked: "It is wrong to suppose that if you can't measure it, you can't manage it—a costly myth."[11] Or as sociologist Bruce Cameron said, "Not everything that can be counted counts, and not everything that counts can be counted."[12]

This book shows you how to use metrics that are subtle and directionally right so that you can successfully experiment with new digital technologies as part of transforming your business.

These traps, by themselves, are not unique to digitization, but they contribute to the continuation of the status quo. In fact, one or more of these success traps very likely applies in your company and may be preventing you from taking the necessary steps towards transforming your business. Recognizing the situation is an important first step; rewriting your company's rulebook is the second step towards reinventing your organization's relevance in the future. And the Digital Matrix will help you develop your perspective about digitization, overcome the success traps, and adapt your business model for this new reality.

YOUR BUSINESS IS BECOMING DIGITAL

If I asked you now, "Is your business becoming digital?" chances are you might say yes. Remember, the question is not about whether you have already implemented a lot of digital changes in your own company. It's more about whether your business is influenced and affected by different digital technologies already or will be in the near future. Specifically, your company is becoming digital if:

- Big data, analytics, and artificial intelligence affect your business processes and how you make key decisions.
- The social web shapes your customers' actions, interactions, and consumptions.
- Mobile apps and cloud computing are essential to how you deliver your services to individual consumers and enterprise customers.
- The Internet of Things (IoT) links all of your products through sensors and software to the broader machine web and the cloud.
- Robotics, drones, and 3D printing are key drivers of the evolution of your supply chain.
- Cognitive computing algorithms and robotics influence how you think of reinventing your business for the future.

Now, go ahead and count how many of the above characteristics apply to you. At least one of them should apply to every company, including yours. And I am also certain that many more will apply in the very near future. As we approach 2025, if not sooner, you, along with most companies, will find digitization at the core of everything your business does.

Let's suppose that you said no to all of the above characteristics. Do you have competitors for whom the answer is yes? Why do they see things so differently from you? How might that give them an advantage over you? And wouldn't *you* rather have that edge?

It's easy to see Google, Apple, Samsung, and IBM as digital companies, because they deliver digital products and services. But so are John Deere and BMW, which are reimagining their products as part of digital ecosystems in agriculture and transportation respectively. It's straightforward to say that Tesla, Uber, Twitter, and Airbnb are digital companies, as their products and services are enabled through digital devices and interactions. But by the end of this book, you will feel comfortable looking at companies such as GE, Philips, Whirlpool, Rockwell, and Bosch not as industrial companies but as digital companies, too. And you will begin to accept that pharmaceutical companies such as Merck, Pfizer, and Novartis are delivering greater value to patient health and wellness as they enhance their traditional competencies in medical sciences with data on how their drugs interact with individual patients. So every industry faces its own digital future, and by extension, every company in it does as well.

Today, we are at an inflection point where old definitions of industries, competition, and organizations do not make much sense, but we do not yet have new ways to define and demarcate pockets of value created by digital technologies. The old rules of strategy and management—experimented and perfected in the industrial age—do not appear useful, yet we do not have new rules of management or value creation and capture. We see companies born digital in the postindustrial age emerge with principles and practices of management that are very different from the companies born before them.

I have created the Digital Matrix as a management framework, to help you understand the forces that are likely to influence the landscape you will encounter in the very near future. I discuss this digital transformation from the vantage point of an incumbent, an established company that has grown up in the traditional, industrial age. This company is at a critical crossroads, and you may well be working in one now. I do not, however, distinguish between functions or levels within companies, because you can play a critical role in the digital transformation of your company no matter what your title or function.

You know that you have to transform, yet you do not know how and when to abandon the tried-and-tested models and practices. In Part 1, I will explain how the Digital Matrix works, introduce the three types of players that cooperate and compete with you in the new landscape defined by ecosystems, and look at the different phases of digital transformation. In Part 2, I will expand on the three phases in more detail so that you will begin to see where your industry and company are located and understand the opportunities and pressures from digital shifts. In Part 3, I will show you three winning moves, starting with how to navigate in dynamic ecosystems, how to work with different companies to co-create new capabilities to deliver new business value, and finally, how to design your organization to reflect the powerful new intersection between humans and machines. These moves are a departure from industrial-age principles rooted in mechanical engineering and reflect new management principles based on computer science. And Part 4 will give you the chance to map your own strategy for the digital future by weaving together nine rules that form the basis of your company's Rules Matrix.

Change is not easy, but it is inevitable, and it's necessary for growth and survival. Whether or not you're a *Fortune* 500 company, do you feel confident about your company's continued success? Do you see signs of the success traps I discussed earlier? Does your organization have the mindset and the skill set to deal with digitization? No matter where you are with digital transformation, with the Digital Matrix (and its resulting Rules Matrix), you have a systematic guide to understand the process and its new rules, evaluate your options, and make rigorous decisions not only in the short term but also for the long-term success of your company. This book is your lens through which to make sense of the shifts and your guide on this important journey. Embrace your role as a leader. Marshal your colleagues to recognize and react to digitization. The success of your company during this pivotal period depends on it.

PART 1

YOUR BUSINESS IS ALREADY DIGITAL

CHAPTER 1

◦━━━━━━━◦

THE NEXUS OF SCALE, SCOPE, SPEED

"**M**OST CARS DON'T improve over time. By contrast, Model S gets faster, smarter, and better as time passes... [It] actually improves while you sleep. When you wake up, added functionality, enhanced performance, and improved user experience make you feel like you are driving a new car. We want to improve cars in ways most people didn't imagine possible."[1]

If this sounds like science fiction, it's not. Tesla Motors has been pioneering electric vehicles and over-the-air software updates. Unlike Alphabet, Apple, Amazon, and Facebook, Tesla Motors may not yet have made the *Fortune 500* list. But Tesla, like the other four, is part of a new breed of companies born in the digital era or shaped by digital shifts that are growing by enhancing their scale and scope of operations at speeds that we have not seen in business ever before. They have already started to exert their influence in your industry and are likely to do so even more in the coming decade.

THE THREE DIMENSIONS OF DIGITAL BUSINESS

In the industrial era, companies expanded their *scale* by increasing sales of their products, which resulted in a higher market share. And this scale expansion was linear based on the company's ability to access the physical, human, and financial capital necessary for growth. Think of Coca-Cola expanding globally and Walmart opening stores across the United States and around the world. Some companies expanded their *scope* by extending their existing product lines and introducing new related ones. Think about Procter & Gamble's steady expansion in household brands, organically and through acquiring other companies, from Ivory body soap to Pampers disposable diapers to Gillette razor blades. Such scope extensions were rather gradual, often requiring significant financial and human capital. This steady and progressive expansion in scale and scope was a successful business strategy.

Digital companies show patterns of scale expansion and scope extension at a *speed* that is wildly different from that of the industrial age. Instead of linear rates of change, digital companies are showing mastery over non-linear, exponential expansion in scale and scope. In doing so, they start to influence industries with new capabilities that take advantage of digital technologies. So you need to locate your business at the nexus of scale, scope, and speed. Here, numbers tell more compelling stories than words.

SCALE

In 1999, Google handled 1 billion search queries; by 2012, that number was 1.2 trillion in a single year and up to 2 trillion searches annually in 2014![2] The company was not in the smartphone business in 2008, but by the end of 2015, more than 1.5 billion devices were running Google-powered Android operating system (os) software.[3] That's Google's non-linear, exponential scale with search and mobile. If you add YouTube, Google has more than 1 billion users (about one-third of the global online audience), and its watch time has grown 50 percent annually for each of the last three years.[4]

Uber, which had a small set of drivers in 2011, had more than 300,000 drivers by the end of 2015, doubling from 150,000 in 2014. On December 30, 2015, it recorded its 1 billionth ride,[5] and six months later it had reached the 2 billion ride mark.[6] By early 2016, it operated in more than four hundred cities in seventy countries. By comparison, yellow taxis in New York City—considered a popular ride-hailing city—recorded about 175 million rides in 2014, according to the *Taxicab Fact Book*. The value of taxi medallions, the highly coveted licenses issued by the Taxi & Limousine Commission, which regulates who can drive a taxi in NYC, is declining.[7]

Perhaps more telling are the comparisons between traditional (what I call "incumbent") companies and their digital counterparts. For example, the long-standing Marriott Hotels had about 760,000 rooms available worldwide in 2015.[8] Airbnb, the online accommodation booking service that started in 2008, had 50,000 listings in 2011, grew tenfold to 550,000 listings in 2014, and increased roughly fourfold to 2 million in more than 190 countries by early 2016.[9]

In retail, Walmart reported 260 million customers in 2015.[10] Amazon, the largest online retailer and Walmart's main competitor, recorded 304 million active customers by the end of 2015, after just twenty years of existence.[11] And, in those same two decades, most of Amazon's e-commerce (dot-com) competitors have withered away or become smaller or niche players. Although several large retailers coexisted in the industrial age based on locational advantage and differentiated merchandise, digital-era retail seems to favor one or two large players alongside countless specialist niche players, a phenomenon that has come to be known as "the long tail."

In the nineteenth and twentieth centuries, railroads and telegraph lowered the costs of transportation and communication, and the successful firms were the ones that built and maintained the organizational capabilities necessary to exploit economies of scale. They invested in the capital equipment necessary for high-volume production and local and global networks of marketing and distribution. And they formalized organizational structures and management systems that allowed them to invest to take advantage of economies of scale. In the words of historian Alfred Chandler, the modern industrial corporation of the twentieth century exploited economies of scale because of the "three-pronged investment in production, distribution, and management."[12]

When I talk about the linear growth of the twentieth century's leading companies, I like to use McDonald's as an example. You may remember the sign "Over [XX] Billion Served" that stood in front of each restaurant and was updated periodically. McDonald's reached the 1 million mark in 1955, the 1 billion mark in 1963, and the 100 billion mark on April 15, 1994.[13] I remember seeing the sign "Over 99 Billion Served" in 1994, and then the restaurant simply adjusted all of the signs to read "Billions and Billions Served" and left it at that. Why? Because McDonald's did not keep track of how many individuals consumed their hamburgers; they merely counted the number of hamburger patties shipped to all their locations.

That's the difference now. Digital-era businesses such as Uber, Airbnb, Netflix, and Google have amassed detailed data on their operations. Google's database of our search queries is now in the tens of trillions, and the same holds true for its mobile platform. Uber collects data on its 1 billion–plus rides and uses it to fine-tune its operations in ways that the taxi companies of yesterday could never hope to achieve. Netflix knows our preferences for movies in ways that cable and television companies never did or could: they simply were not designed to collect, process, analyze, and interpret such data. Airbnb knows where, when, and how long we stay in ways that the hotel chains do not and cannot. Amazon knows our buying habits in richer detail than Walmart ever has.

The bottom line is that if you are still operating on the assumption that scale means the number of products manufactured or sold (units)—and that selling more units relative to your direct competitors means a higher market share and therefore a lower per-unit cost and higher profitability—you may be at a scale disadvantage in the digital world.

SCOPE

How did Apple go from being one of many companies selling personal computers in 2001 to dominating the music and telecommunications industries by 2011? How did Google parlay its supremacy in search into leading the mobile web (Android) and media web (YouTube) and into automobiles and health care within a single decade? How did Amazon go from being an e-commerce bookseller to being a towering giant on the cloud in just twenty years?

Since the end of the Second World War, many corporations have extended their core business into adjacent areas. For example, meat-packing firms took advantage of by-products in their production processes to make leather, soaps, and fertilizers. And Honda used its core engine technology to offer motorcycles, automobiles, lawn mowers, and aircraft engines. By the end of the twentieth century, General Electric (today's GE) was selling not only the home appliances for which it was known but also unrelated products and services such as aircraft engines, entertainment (as part owner of NBCUniversal), and financial services. At that time, most companies that diversified too far from their core markets, such as United Technologies, which was founded as an aircraft manufacturer, and ITT Corporation, which was originally a telecommunications company, were brutally punished by the stock market, and they returned to their core competencies and divested non-core businesses and processes.

Whereas companies traditionally expanded their scope incrementally and relatively methodically by testing and then extending their core competencies in new geographies or market segments or by gradually adding products and services to their core offering (as the auto companies did with financial services, insurance, telematics, and so on), digital-era companies rely on their core competency—data and analytics—to predict with a high degree of accuracy what their consumers want. With machine learning and artificial intelligence, Google, Uber, Netflix, Airbnb, and Amazon can take huge volumes of data and sift it, sort it, and analyze it to expand their scope with new products and markets—even in unrelated industries. And because they can track when, where, and how consumers are reacting—and adjust the offering quickly—they can mitigate risks and capitalize on successes to fuel exponential growth.

So again, the bottom line is that if you are still operating on the assumption that scope means extending your reach only within your own or adjacent industries—and with products or services related solely to your historical core competency—you may be at a scope disadvantage. And if you think that you are only vulnerable to competition from leaders in adjacent industries, you are looking too narrowly at the landscape.

SPEED

You've almost certainly heard Facebook CEO Mark Zuckerberg's motto: "Move fast and break things... Unless you are breaking things, you are not moving fast enough."[14] And that's the idea of speed in the digital world. It's not about being reckless; it's about continuous improvement and iteration, a culture that Zuckerberg calls the Hacker Way, because "hackers believe that something can always be better, and that nothing is ever complete."[15] Using that same premise, Google develops products in the open, adds features daily or weekly, and closely observes how customers use them. This immediate feedback makes customers trusted co-developers. And Tesla maintains and upgrades its cars through over-the-air software updates,[16] which is just another form of speed as a key attribute of digital businesses.

In the industrial age, companies hastened to lock up physical assets such as land and machinery, as well as access to production and transportation. Traditionally, speed referred to the time it took a company to act (and react) to changes in the specific industry and relative to other competitors within that industry. Writing in the late 1980s, George Stalk at Boston Consulting Group argued that: "Companies that meet the needs of their customers faster than competitors grow faster and are more profitable than others in their industries. We argued that time could be the next decade's most powerful competitive weapon and management tool for US companies."[17] Viewed this way, your speed allowed you to reap first-mover advantage relative to others competing against you within the accepted industry definitions. In other words, your ability to be faster in the market hinged on your organization's own clock speed in areas such as product design and development, manufacturing and supply chain synchronization, and so on. It also depended on your information technology department's ability to speed up the back-office processes—often operating on antiquated systems and legacy infrastructure—to support the development of new products. The slowest part of the interlinked processes defined your speed. As long as your competitors were in a similar state, this did not prove ruinous.

Now, the digital players are dictating the pace of customer service with new services that are enabled and delivered via the cloud and through apps on the mobile phone. You not only need to speed up the back-office processes to compete against your traditional competitors but you also need to calibrate the speed of your delivery to the benchmark set by companies born in the digital era. If you are still operating on the assumption that speed means being the first to move into a new market—rather than the fastest to capitalize on the opportunities—you may be at a speed disadvantage.

THE COMBINATORIAL ADVANTAGES OF SCALE-SCOPE-SPEED

In industrial-age companies, scale, scope, and speed acted independently. The scale decisions were handled within individual business units, which first sought to become efficient in production or distribution at the minimum viable scale before expanding based on the available resources, organic growth, and acquisitions. The scope decisions concerned corporate strategy and often involved mergers, acquisitions, and joint ventures, in addition to significantly realigning the resources of existing businesses. Speed often reflected speed to market (first mover versus fast follower) and defined a company as either slow or fast relative to other competitors within specific industries. As an incumbent in a traditional industry, you already know how to tap into the advantages of scale, scope, and speed within your industry. You may have developed an advantage in one or more of these dimensions compared to your traditional competitors.

As your industry digitizes, progressively in some cases and rapidly in others, you need to look at these three dimensions of your business as being interconnected. Scale and scope still define your company's strategic ambition and address the question: What set of businesses should we operate and at what scale? However, scale at speed creates not first-mover advantage but *fast-mover* advantage, which may currently be limited by your company's internal organizational processes and systems, if they cannot recognize and respond to the shifts as quickly as some of the newer companies. Changing scope at speed also reflects fast-mover advantage, where the advantage may lie not necessarily in launching products but in tapping into scarce critical resources such as unique interconnected data, patents, talents, or research and development projects, often executed with others.

How well you stack up against not only other incumbents, who themselves are transforming, but also against newer-age companies that are aiming to disrupt and transform your industry may well define your ability to compete and win in the digital realm. Those companies that take maximum advantage of scale, scope, and speed together are able to gain significant advantage in the digital business world. First, with data and analytics and connectivity, you can now extend your footprint beyond your core firm's boundaries and tap into extended ecosystems. Second, through sensors, software, and connectivity, you now have the capability to collect data, process information, and learn in ways that would have been difficult if not impossible in the industrial world.

The ecosystem advantage

Whereas scale advantage arose in the industrial world from assets that a single firm controlled and the units that it produced, in the digital world, scale advantage comes from being part of an ecosystem that includes key partners that play complementary roles. Ford and GM's scale depends on the number of cars produced by them, but Uber's scale is defined by the number of cars it has in its network on a global basis as well as locally in every one of the four hundred–plus cities in which it operates. Whereas Nokia's scale depended on the number of feature phones it manufactured and sold globally, Google's scale advantage, as the architect of the Android mobile operating system, depends on the number of devices produced by its hardware partners in the ecosystem and the number of software apps written by the developers for its operating system. In the industrial age, scale is the result of what a firm does by itself using the assets that it controls and the units it produces. In the digital world, scale is the result of what it may produce by itself *plus* what it can achieve with its partners in the ecosystem. *Tap into the scale advantage conferred by your ecosystems.*

Just like scale, scope advantage in the digital world comes through being part of an ecosystem. You may wonder what's the contrast here between industrial and digital: in industrial, the relationship between a company's core area and its adjacencies had to be pretty close for customers to accept the link; in digital, data as a core area is infinitely malleable so that companies that collect data can more easily apply it across a wide range of platforms, as in mobile platforms. With their core mobile software—Apple's ios and Google's Android—digital giants can logically extend their scope with different apps. Payment apps, such as Apple Pay and Android Pay, supported by merchants and global retail banks create an ecosystem that allows Apple and Google's parent company, Alphabet, to move into the seemingly unrelated area of retail finance. But they do so for different reasons—Apple to enhance the use of its phone and watch but explicitly not using the information on such transactions, and Google to better target its advertisements by using that information. *Use the scope advantage of your ecosystems.*

In contrast to the industrial world, where a single company could gain an advantage by being the first one into a new market, in the digital world, everyone in an ecosystem has to move at more or less the same speed. Since not all the competencies lie inside your firms, you have to rely on the ecosystems. It's like being on a relay team: one super-fast runner is not going to win the race for the group, though one super-slow runner, like one super-slow company, could lose the game for the entire ecosystem. In other words, your critical skills and capabilities might increase your chances of joining an ecosystem, but your ability to stay up to speed (or even enhance the speed) could well be the deciding factor. Sony PlayStation has succeeded over the past decade because it has mobilized its game development partners with the pace of successive console developments. *Structure your relationships to profit from the speed advantage of your ecosystem.*

Throughout this book, we will explore in more detail how you connect to different ecosystems to gain such advantages to craft your winning strategies.

The learning advantage

An important characteristic of scale-scope-speed is learning from products and services in use and adapting their characteristics to the specific needs of individuals. So how do we think about collecting data? Whereas companies in the industrial world collected data about a few attributes, mostly focused on operational efficiency, and analyzed this coarse, aggregated data over time, digital companies are constantly recording data with detailed attributes and analyzing it using new tools to discern patterns of preference and fine-tune their strategies. To determine how many burgers it sold, McDonald's counted the total number of hamburger patties shipped to its locations. In contrast, Starbucks uses its apps and loyalty programs to understand not only how much coffee it sells but when and where its customers buy their coffee, how they prefer it, how much they spend per transaction, and so on.

Products behave differently under various conditions, and no amount of testing in the lab is enough to understand the particular behavior in actual conditions of use—whether they are tractors on the field or aircraft engines in flight or cars on the road or washing machines in the home. Now, companies monitor their products in so many different locations (even remotely) at scale and in near real time that they have more opportunities than ever before to learn about them, modify them, and even correct mistakes before the impact is felt too widely. *Learn from products in use at scale to glean early warning signals.*

Companies in the industrial era expanded their scope by branching out to related products or markets, and they made these decisions based on pre-established patterns followed by others and based on analyzing data from market research and other coarse data. In the digital age, companies can actually predict areas of inefficiency by using analytic software and expand into seemingly unrelated areas. For example, GE, after borrowing from the playbooks of Apple, Google, Microsoft, and others, is now on a new mission to use software, apps, and data plus analytics in four industries: buildings, power, industrial transportation, and health care. The company's Predix platform with analytics as the foundation allows it to predict areas of major inefficiencies within and across diverse industries and solve them better than anyone else, including their own customers.[18] Furthermore, we can now not only collect data on our own products but we can also see how products from different companies operate together to solve customer problems. For example, in a health care setting, firms can monitor how their device or medication interacts with other treatments across a wide spectrum of different patients. Bearing in mind proper safeguards for privacy and security, all the companies contributing products (and services) can learn from the data and tailor their products to individual patients, specific treatment plans, and/or any number of other variables. Similarly, companies such as Amazon, Google, and Facebook have access to troves of customer data that could be mined for learning advantage. *Learn from customers that use complementary products to proactively improve key features.*

Industrial-age companies spent a lot of time before starting experiments to make sure that the goals were well specified and the mandates well established. Digital-age companies start projects on the backs of passionate people who try, hack, fail or succeed, learn, and adapt. They "fail fast" and "pivot,"[19] which simply means they learn fast with data, adapt their prototypes, and reflect on customer feedback. They pivot along different dimensions, such as customer segments, channels, revenue streams, partnerships, and value propositions. Since every interaction is an opportunity to collect data about the products and systems in use, they move fast to embrace new ideas not because they are slavish but just to learn at a deeper level. The ultimate advantage at the nexus of scale-scope-speed, then, is reflected in learning through experimentation and taking advantage of the greater scale and scope of your ecosystems. For example, Netflix used machine learning, analytics, and A/B testing—comparing two different versions of an offering—to create its personalized video recommendations.[20] Doing so at speed—understanding the validity of how your assumptions operate and iterating fast based on the results—allows you to refine your working hypothesis in key areas. Or as Eric Ries, an expert on lean startups preaches, validate your learning scientifically "by running experiments... to test each element of the vision," and "build-measure-learn" to accelerate your feedback loop.[21] *Learn from experimentation through data and analytics.*

MASTERING YOUR EXPONENTIAL TRAJECTORY

Learning from ecosystems is continuous. As ecosystems help you scale further, you gain more opportunities to learn. As ecosystems help you expand the scope of your business footprint, so too do your learning opportunities expand. And as you extend your scale and scope at a faster speed, you increase your learning opportunities further. So, scale, scope, and speed are mutually reinforcing. What emerges at the nexus of scale-scope-speed is a new focus on a *non-linear, exponential trajectory,* and your ability to master these shifts as your industry digitizes and evolve exponentially is an important new strategic requirement.

To understand this non-linearity, let's look at an example. Ray Kurzweil, author and resident futurist at Alphabet, believes in a "law of accelerating returns"[22] arising from the exponential increase in the power and functionality of personal computers and smartphones. He traces the ever-quickening evolution from "the mechanical calculating devices used in the 1890 US Census, to Turing's relay-based machine that cracked the Nazi enigma code [in 1937], to the vacuum tube computer that predicted Eisenhower's [presidential] win in 1952, to the transistor-based machines used in the first space launches [in the 1960s], to the integrated-circuit-based personal computer [in the 1980s]" over the last 110 years. Looking ahead, this exponential increase in the functionality of computing power will extend to other areas such as devices connected to the Internet of Things, wearable computing devices embedded into clothing and footwear, health care devices, drones, 3D printers, robots, and automobiles. As the number of such powerful network-enabled devices increases to 50 billion or more over the next decade,[23] managing the exponential shifts in digital business will become the top priority. Non-linearity in technical features and performance improvements may be obvious to the technologists on your team, but your job is to recognize and respond to the opportunities and threats in this new business landscape of cross-industry ecosystems and extended social and professional networks. The Digital Matrix, which we will look at next, is designed to help you do just that.

CHAPTER 2

THE DIGITAL MATRIX

IMAGINE YOU HAVE entered a room to play a new business game. You are excited but also apprehensive because you haven't played this game before. You don't know the rules. You don't know all the opponents. You know neither their skills nor ambitions. You may recognize some of the players from a past game with different rules, including a few who have supported you in the past but whose current roles and motives you do not discern. You do not comprehend the total number of contestants. What you do understand is that this game is not purely competitive: you can partner with other individual players to form coalitions that can play against other coalitions of players. You also recognize that the game is played over time, and that new players enter at different times and form new relationships, and that some existing relationships get solidified and new relationships are formed. Some understand linear progression, whereas others are masters at exponential trajectories. The players develop new capabilities as they progress, and the payoffs get bigger and stronger, which means that the losses also get sharper and more severe. No one has been crowned as a grand master yet, and everyone believes that they have a fair chance of winning. Your company is in line to play this new business game in the digital future. You're ready, aren't you?

THE DIGITAL MATRIX AS YOUR LENS

To help you understand this new game, decipher the rules, and develop your heuristics to win, I've developed the Digital Matrix. It's a framework that allows you to see how three types of *players* (including you) use a variety of digital technologies to shape the future of your industry and influence your company's strategic actions and responses over three distinct *phases* of digital transformation, in which these players adapt and design effective business models for the future using three winning moves. Picture the Digital Matrix as a control panel with nine screens arranged in a three-by-three grid. Along one axis are the three types of players. Along the other axis are the three phases of digital transformation. In this chapter, I'll identify these players and introduce each of the phases to show you how the framework can help you make sense of the game.

Remember that the Digital Matrix is a management framework; it's not a technical or tactical one. It invites discussion and debate, raises options and investment trade-offs. It's a call for you to let go of some successful, but now outdated, past practices to embrace new rules and continually experiment with new approaches and adapt them to suit your needs. Three main characteristics set the Digital Matrix apart from other books and approaches.

It's not a matrix of technologies

The forces of digitization are not simply a set of technologies. The Digital Matrix recognizes that you are developing your strategy against the broader landscape of decisions and sequenced actions taken by three sets of players, and that these players, not the power and functionality of technologies, drive the transformation. It takes into account that companies in different roles and phases are all embracing, experimenting with, and exploiting digital technology to craft new business logics that give them some advantages, but they are absorbing and assimilating it differently to create new capabilities, establish new relationships, and seek differentiated drivers of value.

The Digital Matrix captures the core actions by different types of play-ers (business-pull) rather than focusing on the devices that drive them (technology-push).

It's not static

Digital strategy is not a set of specific actions carried out in a particular order at specific stages of a predefined life cycle or "hype cycle"[1] of tech-nologies. The Digital Matrix recognizes that companies and industries are constantly moving through three phases of transformation that evolve quickly, and therefore, it does not try to impose a one-size-fits-all solution. It takes into account that the players, the technologies, the moves are continuously changing and creating new conditions.

The Digital Matrix brings out the dynamics—the actions, reactions, and the follow-on decisions—to ensure that you are continually taking advan-tage of developments with different technologies to evolve your business for the digital future.

It's not one-dimensional

Success depends on being able to focus on three players and three phases simultaneously. Isolating any one variable is not enough, and it's this idea that is the crux of the Digital Matrix. Once you realize that your thinking must transcend traditional boundaries, that you must comprehensively examine not only your own actions but those of oth-ers in the game, then you will see what's happening in other settings, better understand how to set the rules of the game, and be in a position to create the strategic alliances and big moves necessary to exploit new opportunities and win big.

The Digital Matrix sets out the multidimensionality through the nine screens, such that your actions in every phase—which are deeply interconnected and mutually reinforcing—invite reactions and responses from the three sets of players.

As we begin, know that you have to prepare for this strategy game differently. Forget about digital as one specific technology. Discard piecemeal solutions such as revamping your information technology department or appointing a new chief digital officer or establishing a strategic task force to look at new digital technologies and recommend possible acquisitions and alliances. Instead, start thinking holistically about how digitization could drive the design of your business strategies and the structure of your organization. Consider how the different forms and functionality provided by digital technologies could provide better value for customers of your products and services. Recognize that the real payoffs happen when you can access and analyze data across functions and across companies, and then make changes to the products, processes, and services that connect all of them.

This transformation is about systematically recognizing how digital technologies—the ones that we have seen, the ones that are on the horizon, and the ones that may still be in development in research labs and universities—begin to change the scale, scope, and speed of your business. They begin to shape your strategy in unprecedented ways. You need to look at the entire business system, and that means identifying and analyzing all of the players that intersect with you in this digital strategy game—represented across the nine screens.

THE THREE SETS OF PLAYERS

In the digital world, you will find yourself contesting against a broader set of players than the ones you may be familiar with today.

PLAYER 1: INDUSTRY INCUMBENTS

You know your current competitors, what I call *industry incumbents*, well. Chances are that you have scanned, analyzed, and understood how these historical, traditional competitors in your industry respond to market shifts. You recognize them well enough to anticipate their likely actions and you have a repertoire of responses to these. Soon, however, you are likely to be positioned in networks comprised of not only traditional competitors you know well but also newer ones you don't. I want to particularly highlight two additional sets of players.

PLAYER 2: TECH ENTREPRENEURS

Ambitious, upstart *tech entrepreneurs* have brazen and auda-
cious views on how they can disrupt and reorder the business
world. These are PayPal in financial services or Tesla in automotive or
Palantir in analytics and security. Or China's Xiaomi in consumer electronics (trying to outdo Apple with beautifully designed digital devices)
or Didi in the ride-sharing sector (taking on Uber but with global ambitions) or India's Flipkart in e-commerce (aspiring to become the answer
to Amazon). These companies and others are born digital with a blatant
disregard for management rules from the industrial age. They believe
in crafting business models that promise to deliver unparalleled value
to customers by using the power of digital technologies. Algorithms
and automation guide their thinking, data becomes their differential
resource, and analytics their distinctive competency. They think beyond
narrowly defined industry boundaries and evolve their business models
by taking advantage of different digital technologies. Not all upstarts
will succeed, we know that, but those few that do could turn out to be
your formidable future competitors or trusted partners as you transform.

PLAYER 3: DIGITAL GIANTS

The third category of player I call *digital giants*. These companies, including Alphabet, Amazon, Apple, Facebook, IBM,
Microsoft, and Samsung, have progressively extended their influence
beyond their traditional industry into yours. They were yesterday's
tech entrepreneurs that have now grown up and extended their scope
into industries in which they previously only supplied technologies or
managed back-office operations. Their core business model may still be
to deliver digital products and services, but some of these giants have
increasingly partnered with industry incumbents to help those companies transform their business models for the digital age.[2] The end result
for such giants has been more vertical and horizontal integration into
certain industries—including yours.

You and your fellow incumbents in your industry will be in ecosystems that include tech entrepreneurs and digital giants. What you'll

discover is that in such ecosystems, not all relationships are competitive and adversarial. Some are cooperative and others are competitive. I will discuss this more in Chapter 6. In this new business landscape, relationships that are cooperative in one time period may become competitive in another time period, and vice versa. And, increasingly, some of these relationships will become simultaneously cooperative *and* competitive, a situation I call coopetitive. I will discuss this more in Chapter 7. So the digital business ecosystem is a fast-changing playground. You need to understand the different viewpoints and capabilities of each of these players in order to know how best to interconnect your company with them.

Let's look at the three sets of players in more detail, using the global automotive industry as an example.

The three sets of players in the digital auto industry of tomorrow

The industry incumbents include GM, Ford, Toyota, BMW, Mercedes-Benz, and other traditional automakers. But think about how tech entrepreneurs such as Uber, Lyft, and Tesla, and digital giants such as Apple and Alphabet, have either partnered with some of the incumbents or introduced their own innovations. Already we've seen how digitization has influenced product architecture (hybrid and electric drivetrains that complement gasoline-driven cars), design (rapid prototyping using technologies including 3D printing), manufacturing (modular platform assembly capable of different permutations and combinations for different segments), service enhancements (telematics for in-car navigation, safety, entertainment, and communication that are updated over cellular networks), and business models (traditional ownership challenged by transportation-as-service).

When viewed this way, it's also easy to see that pervasive digitization extends to the broader business arena that includes manufacturers of automotive subsystems too, such as Dana Incorporated, which supplies power train components such as axles and drive shafts; Continental and Firestone, which provide tires; Bosch, which makes auto electronics; and countless dealers and service stations. In other words, you can no longer look at manufacturing, supply chain management, design, and delivery of industrial combustion engines as individual activities; they are inextricably linked in new digital business ecosystems. Every single participant—every incumbent, tech entrepreneur, and digital giant—must assess how it could be relevant both independently and working with other players during the transformation of the automotive sector.

Which tech entrepreneurs are or should be on the radar of these auto industry incumbents? Tesla Motors, certainly. Elon Musk, the legendary entrepreneur of PayPal fame and now the CEO and the visionary architect of Tesla, has masterminded the evolution of an electric car and made the company's trove of patents available to others to accelerate the electrification of the entire transportation system. Tesla, which was market capitalized at more than 50 percent of GM and Ford in mid-2016, is a credible new competitor.[3] When the company unveiled its Model 3 in March 2016, more than 300,000 would-be buyers placed preorders with a cash deposit of $1,000 each within a week. GM and Ford are #8 and #9 on the *Fortune* 500 list for 2015. Tesla is a tier below at #588. Could Tesla enter and rapidly ascend this leaderboard over the next few years? The odds are high indeed if its Model 3 succeeds.

In addition to Tesla, other niche companies are worth noting. A San Francisco startup named Automatic Labs[4] makes an adapter that can be plugged in under the steering wheel and, with an accompanying app, provides detailed data about the car's mileage, gas usage, engine health, and driving performance. The distinctive feature is that car owners can install it as an add-on without having to provide any data on driving patterns back to the manufacturers. Furthermore, the company seeks to develop the "automotive cloud" to enable a suite of services including cloud-based auto insurance and intelligent car care services. Peloton Technology[5] connects multiple trucks through a vehicle-to-vehicle communication system (truck platooning) that controls their acceleration and braking systems virtually to reduce accidents and enhance efficiency. And, of course, Uber and Lyft are two other new-era companies that are at the forefront of providing drivers on demand now and could provide drive-on-demand without relying on drivers tomorrow in select cases. Although Uber may be the most prominent venture-backed disruption in the automobile sector (valued in the $60 billion–plus range in June 2016[6]), incumbents such as GM, for instance, have invested significantly in Lyft[7] ($500 million in January 2016) as part of a long-term strategic alliance to create an integrated network of on-demand autonomous vehicles in the United States. Other automakers are exploring different service-based experiments that I will examine throughout the course of the book.

Most tech entrepreneurs start out with deep, focused expertise and specialized skills in specific *vertical* areas, such as automotive (auto tech), health care (health tech), retail (retail tech), agriculture and farming (farm tech), or financial services (fin tech). So depending on your industry setting, you must recognize the potential role of such specialized entrepreneurs. This is in contrast to the digital giants that have *horizontal* expertise—that extend their core distinctive expertise across different scope domains as we saw in Chapter 1.

What do horizontal scope extensions look like here? We have already seen Alphabet and Apple signal their intent to be key influencers as the industry digitizes and transforms. Alphabet's Android Auto and Apple's CarPlay are software programs that allow for consumers (drivers) to seamlessly extend their smartphone experience into cars. And it's safe to extrapolate that these two digital giants want to delve deeper into automobile architecture, especially with software-driven functionality such as automatic parking, piloted driving, and fully automatic driving on highways. Fiat Chrysler entered into a preliminary agreement with Alphabet in May 2016 to co-create one hundred prototype self-driving minivans in which the digital giant tests its self-driving technology.[8] We can only surmise that the results from this initial experiment could lead to a deeper relationship between the two companies or lead them to pursue different avenues. Toyota has invested in Uber to explore ride-sharing collaboration and to establish new in-car apps and services,[9] Volkswagen has invested $300 million in Gett,[10] a German cab-hailing startup, and several more linkages can be expected in the coming years.

The lesson here is that you, like GM, Ford, Chrysler, and the other incumbent companies in the automotive sector, cannot effectively develop a digital strategy without first explicitly analyzing how the other incumbents in your industry are crafting their own digital transformation moves. Your industry competitors could be acquiring digital capabilities and absorbing them into their core organizations or forming alliances. Tech entrepreneurs could be bringing a new mind-set and skill set to solve the core business problems of your industry's customers. So the reference set for digital strategy expands beyond the industry incumbents and includes digital giants, which horizontally extend their competencies into your industry, and tech entrepreneurs, which bring vertical, specialized expertise to solve problems in new and effective ways. And the same three types of players operate and influence the trajectories of transformation in other sectors, as we will see later in this book.

These three types of players exist on one axis of the Digital Matrix. Now, let's focus on the other axis.

THE THREE PHASES OF TRANSFORMATION

Industries change. They grow, they shrink, they transform. That's not new. In the digital age, however, these transformations occur much more rapidly, and they are no longer linear or chronological. Incumbents must simultaneously manage their actions and interactions with all three sets of players through three phases of evolution that exist at every point in time.

PHASE 1: EXPERIMENTATION AT THE EDGE

There's the embryonic phase, during which experiments with digitization happen and evolve. As one digital idea matures, another experiment emerges, such that digital business experiments are happening all the time. I call this phase *experimentation at the edge*. This is when lots of ideas—some far-fetched, incredible-sounding ones and other more realistic and potentially valuable ones—go from sketches on napkins and slide decks to prototypes, pilots, and products.

During this phase, you could already be focused on the experiments that you're engaged in to adapt your business model. That's necessary, and you should focus on them. But the thrust of this phase is looking at the landscape of experimentation undertaken by a wide range of firms, even those beyond your immediate industry boundaries, to make sense of their potential implications to your industry's ways of working. One prominent area with digitization is the shift from individual products to interdependent products linked via *platforms*.[11] In essence, a platform is a base on which others can build a business. Think of a computer operating system, like Apple's iOS or Google's Android. It is multi-sided in that it connects many different types of companies transacting with many different types of customers and gives them a standardized way to get paid or exchange value of different kinds.

Does this trend complement and enhance your current business model or could it fundamentally transform and disrupt it? When do platforms arise and how could they affect you? Even if platforms have not emerged in your specific setting, you have to understand the conditions that could give rise to platform-induced competition and transformation: the hotel industry executives were not expecting platforms to upend their business logic until Airbnb showed the way. Beyond platforms, your requirement in this phase is to make sense of the signals at the periphery of your current industry and pre-established business models that could challenge the fundamental assumptions that underlie how your business earns revenue and makes profits.

Making sense of experiments at the edge involves deeper thinking. In 2005, when Apple's chief executive officer Steve Jobs announced: "We've worked closely with Motorola to deliver the world's best music experience on a mobile phone,"[12] the first-level interpretation would have been that Apple's iTunes software and business model had been ported from standalone iPod devices to Motorola's phone. For those in the music industry, it signaled the deeper influence of Apple. But what about the second-order implications for companies beyond the music industry—say, the telecom industry that was dominated by Motorola, BlackBerry (RIM), Nokia, and Sony Ericsson? Executives in those companies should have asked what Apple could do with iTunes that was different and significantly better than the Symbian or BlackBerry software programs running on their mobile phones at the time. Could Apple's experiment with Motorola in 2005 lead it to enter the mobile telephony industry?

The deeper interpretation of this experiment at the edge should raise follow-on questions such as: What else is needed technically for Apple to launch its own mobile phone? If music is one application that's readily suited for mobile phones, what other applications are possible and probable with future developments in software and user interface? Such questions must lead to subsequent richer investigations and analysis with the right mix of management and technology competencies to connect the dots and compute the probabilities of plausible innovations.

More recently, Paul Taylor—an ex–Google employee—has founded a new company called ThoughtMachine. In July 2016, it announced Vault OS, an operating system that creates "banks that run in the Cloud... It uses a centralised, permissioned cryptographic ledger as a single source of truth for all transactions. This ensures the highest degree of security... and all banking products use a system of smart contracts."[13] Is this just brazen ambition from an upstart or could it upend the financial services industry? As outlandish as Vault OS's claims may seem at this time, it's worth considering what else may be needed to make this vision a reality. Who else might work with ThoughtMachine's Vault OS to propel this idea forward and radically restructure the banking industry as we know it now? Or who else could take advantage of the blockchain technology on which Vault OS is based to create another viable new approach in the financial services industry or elsewhere? Only once you've considered these questions should you declare this vision unworkable or unworthy of consideration for your own experimentation.

Take this same approach to deeply interpreting the possible outcomes when you look at other areas—drones from Amazon, Alphabet, and Facebook; virtual reality headsets from Microsoft and Facebook; chatbots from Apple, Amazon, Microsoft, Alphabet, and so on. Think through what drones could do beyond providing logistics support to farming, mining, and disaster assistance as AT&T tries to explore their use not just to inspect cell towers but also to provide coverage during time-specific high-demand situations such as concerts or sporting events. Imagine virtual reality beyond gaming in areas such as education and training in specialized areas of health care. Consider the second-order privacy and identity implications of chatbots such as Amazon's Alexa and Apple's Siri. When experiments mature from labs and research organizations to the mainstream, they become ready for prime time—to serve as the foundation of newer, innovative business models explored by any of the three types of players involved in the digital business game. That shift gets us to phase 2.

PHASE 2: COLLISION AT THE CORE

In phase 2 of transformation, ideas have evolved from prototypes to business options. Experiments have shown the difference between possibilities (Could it work?) and probabilities (How do we make it work so it's profitable?). This phase sees genuine tension between established ways perfected in the industrial age and new digital ways experimented just recently. I label this phase *collision at the core*, where digital rules challenge traditional industry practices and pre-established rules of engagement.

This collision seems to happen gradually in some industries and much more rapidly in other settings. Even within a single industry, some companies may face the intensity differently from others based on their unique strategic differentiation. And the pace of change can accelerate suddenly. For example, at the end of the twentieth century, e-commerce was at the experimentation phase and Amazon was an ambitious upstart, along with many companies that promised to revolutionize retailing. In 2000, Walmart didn't feel the heat from Amazon, or any of those companies. It considered e-commerce simply an additional channel to its core physical stores and nothing more. It saw online sales as an extension of its marketing strategy, an occasional push on days like Cyber Monday (the Monday after the US Thanksgiving holiday weekend), to encourage customers to use the fast Internet on their office computers to make purchases and boost end-of-year retail sales. It didn't anticipate that the global logistics infrastructure would grow as robust as it has become today.

By June 2016, Amazon's market capitalization was higher than Walmart's,[14] the world's largest retailer in terms of sales and employment. Senior executives at Walmart and other major big-box retailers have clearly noticed that the collision at the core of the retail industry between brick-and-mortar and e-commerce business models is becoming more severe. Amazon might have gunned for brick-and-mortar booksellers in 1994, when it started out, but that was just the beginning. As we saw in Chapter 1, Amazon's mastery over scale and scope at speed is what differentiates it from many other e-commerce companies of the early twenty-first century. In the last decade, Amazon's business model has upended the established structure and practices of the wholesale and retail landscapes. And looking beyond the US, Alibaba has not only emerged as China's digital retail juggernaut but also clearly has more global ambitions.[15]

Collision at the core occurs because digital technologies make an impact in two ways. One is that digital products and services challenge traditional products and services. Traditional analog wristwatches vie against smartwatches and digital wristbands. Conventional standalone refrigerators from GE, Bosch, and LG Electronics compete against those connected to the Internet of Things (IoT) network by Samsung and others. As more products become digital, this collision intensifies, and competitive interactions between traditional industries and the digital giants and ambitious entrepreneurs are exaggerated.

The other is that newer organizational models are based on principles of computer science—automation, algorithms, and analytics with software models—rather than mechanical engineering rules that gave credence to scientific management principles proposed by Frederick Taylor.[16] The result is continuous experimentation, dynamic coordination with partners in ecosystems, and machine learning at speed with powerful machines and smart humans working together rather than standardization, specialization, and value-chain optimization. In other words, the collision between industrial-age companies and digital-era companies is a clash of assumptions around value delivery and organizing logic. Those who survive this clash live to influence the new rules, which is the third and final phase.

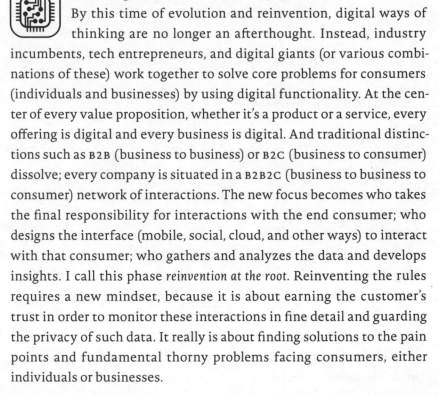 ### PHASE 3: REINVENTION AT THE ROOT
By this time of evolution and reinvention, digital ways of thinking are no longer an afterthought. Instead, industry incumbents, tech entrepreneurs, and digital giants (or various combinations of these) work together to solve core problems for consumers (individuals and businesses) by using digital functionality. At the center of every value proposition, whether it's a product or a service, every offering is digital and every business is digital. And traditional distinctions such as B2B (business to business) or B2C (business to consumer) dissolve; every company is situated in a B2B2C (business to business to consumer) network of interactions. The new focus becomes who takes the final responsibility for interactions with the end consumer; who designs the interface (mobile, social, cloud, and other ways) to interact with that consumer; who gathers and analyzes the data and develops insights. I call this phase *reinvention at the root*. Reinventing the rules requires a new mindset, because it is about earning the customer's trust in order to monitor these interactions in fine detail and guarding the privacy of such data. It really is about finding solutions to the pain points and fundamental thorny problems facing consumers, either individuals or businesses.

Take for example a not-so-distant future scenario in a digital home. Whirlpool has historically been a B2C brand with a B2B process that ended at the retail store, such as Walmart or Best Buy. Although Whirlpool may have more accurate data on the sales per model of its washers and dryers in each channel, its data at the level of individual homes in terms of actual installations may be suspect. But the Internet of Things changes all that. With its washers and dryers connected to the Internet, Whirlpool can track and monitor its products in every home to understand usage patterns at a very detailed level. Digitization allows Whirlpool to track this usage in real time, spot early warning trends and learn from them, carry out focused experiments, and deliver a much higher level of customer service than it could in an unconnected world. Whirlpool's business model shifts from delivering its product to a retailer for distribution to guaranteeing the performance of its services to consumers through its networked products. The more consumers that connect, and the more products they connect, the better Whirlpool can fine-tune its information and help make homes the smart hubs of people's lives.[17]

Reinvention at the root is about strengthening the intuition and judgment that have been valid modes to arrive at decisions in the past with tangible support from data and analysis. Developing insights and applying them faster and better than competitors is a key to success in this phase, so agility and the ability to adapt quickly are essential. To be fast and nimble, companies use data and analytics but, more importantly, shed the bureaucracies that have been characteristics of the industrial world. GE's chief executive officer Jeff Immelt observed in an interview: "Industrial companies are in the information business whether they want to be or not ... We want to treat analytics like it's as core to the company over the next 20 years as material science has been over the past 50 years."[18] He recognized early that for GE to win the digital business game, the company must fuse physics and digitization, link machines and data and act upon information in real time, and become a platform and applications company for different industries,

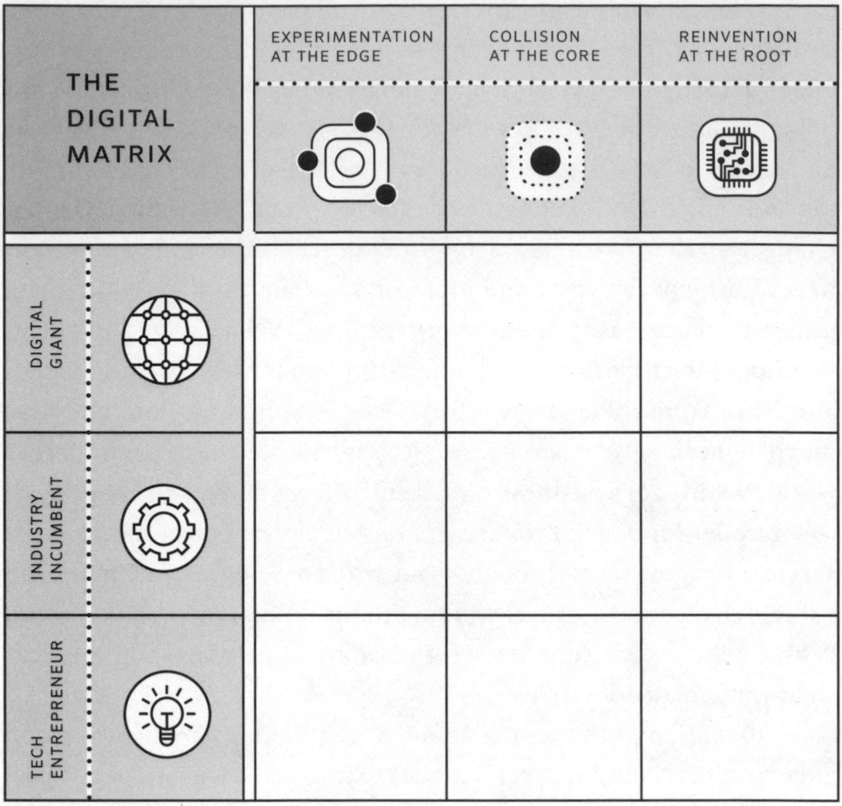

THE DIGITAL MATRIX	EXPERIMENTATION AT THE EDGE	COLLISION AT THE CORE	REINVENTION AT THE ROOT
DIGITAL GIANT			
INDUSTRY INCUMBENT			
TECH ENTREPRENEUR			

FIGURE 1 The Digital Matrix

just as Alphabet and Apple have done for the consumer Internet over the last decade. This is an exciting frontier: for the first time, we are talking about the digitization of the physical web, not just the virtual web for information-based companies. Companies such as GE, Bosch, John Deere, and Whirlpool are leading the way, and the new rules they establish are defining which other companies survive and which get left behind when these new digital business models collide with their old, inflexible, traditional ones.

With your new understanding of the two fundamental axes of the Digital Matrix, the players and the phases, imagine that you are situated in one of the nine screens on the control panel. You know already that the game is not going to be played out on just that one screen; the moves and countermoves taking place in the other screens will influence your moves and countermoves. So let's help you to position yourself on the Digital Matrix. Then you can better understand your position relative to the other players.

YOUR POSITION ON THE DIGITAL MATRIX TODAY

No matter what industry you are in or what service or product you offer, your company is located at the nexus of three players and three phases as shown in Figure 1.

THE DIGITAL MATRIX helps you understand exactly what your position is at this point, where you are situated among the nine screens on the control panel. If you are a tech entrepreneur, your position is along the top three screens; if you are a digital giant, you are along the bottom three. If you are an industry incumbent, you are in the middle set. That's defined by the first criterion—namely, the type of player you are. Now, let's look at where a particular industry is along its evolution.

1. Identify your industry along the three phases.

Is your industry in the experimentation phase, where you can see digital possibilities within your own company and other industry incumbents, or where tech entrepreneurs or digital giants are dabbling with some interesting innovations, but your core business model has not changed? The global agriculture industry is in this phase: experiments with sensors, climate data, and tractors connected to the cloud have been happening in recent years, but the agriculture business models remain fundamentally unchanged. However, precision farming and digital agriculture are on the horizon.

Or is your industry in the collision phase, where upstart entrepreneurs and digital giants are using powerful technology to challenge and disrupt the way you and your traditional competitors do business? The global automotive industry is at this stage, where the industry incumbents are being forced to rethink their business models by upstarts that redefine product architecture and use (e.g., Tesla) and offer transportation as service as an alternative to car ownership (e.g., Uber and Lyft).

Or has your industry reached the reinvention phase, where industry incumbents, tech entrepreneurs, and digital giants are all using powerful technology and new approaches to solve fundamental business problems? The media and entertainment industries are being fundamentally reshaped by companies such as Google (YouTube), Amazon, and Netflix as well as Disney (ABC) and Comcast (NBC), which are monitoring customer preferences and using that data to tailor their offerings to individuals, fine-tune their products and services, and pivot quickly when changes are needed.[19]

Find your industry's current position among the nine screens. If you are an industry incumbent, you can now more precisely situate yourself in one of the three screens of the middle band. But do take the next step to identify the relevant tech entrepreneurs and digital giants that are already or likely to be in your orbit.

2. Assess your relative position against your direct
    competitors and other industry incumbents.
    Companies vary widely in terms of the importance they give to new trends, and digitization is no exception. Are you leading relative to other industry incumbents? Have you, like John Deere or Monsanto in the agriculture industry, recognized the certainty of digitization and taken an early lead over your other industry incumbents by making acquisitions and entering into alliances to gain initial advantage? Do other incumbents look to you, as incumbents in the manufacturing industry look to Siemens or Bosch in Germany as benchmarks, to understand possible directions to follow with digital shifts?

    Or have you, like many firms that think digitization is not yet relevant as a companywide strategic issue, allowed others to lead with digitization? If you have consciously pushed digitization to the back burner because of other pressing challenges, I urge you to keep track of how others in your industry have invested in digital technologies and, more importantly, understand the pattern of shifts in other industries that may provide pointers for you. Or perhaps you are in the middle of the pack, on par with other incumbents that are paying minimal attention to digitization where it's needed but nothing more.

Calibrating your position on the "lead-parity-lag" scale is a useful exercise. Ultimately, it does not matter whether you have already taken any steps relative to other traditional industry incumbents. What matters is that you are prepared to systematically accept digitization as a transformative force in your industry going forward. Knowing your position on the nine-screen grid of the Digital Matrix is a good start. But it's just the beginning. As I've said, no company's an island in the digital world. Value creation and capture involve interlinking islands. So the Digital Matrix recognizes that tech entrepreneurs and digital giants are potential allies but also possible adversaries. The actions and interactions they take among themselves, with your industry incumbents, and with you, profoundly influence your ability to craft winning strategies.

WHY DIGITIZATION MATTERS TO INDUSTRY INCUMBENTS

I've written this book primarily for managers of incumbent companies. Regardless of the phase of digital transformation you and your traditional industry are in, my hope is to break you free of the success traps that may be preventing you from recognizing the power and promise of digital technologies. Having said that, these discussions apply equally to all players—tech entrepreneurs and digital giants included—and to employees at any level within a business. Even if you have not yet thought much about digitization beyond mobile apps and social media presence, you can learn from the experience of others who have dealt with the same types of opportunities, uncertainties, challenges, concerns, apprehensions, and assumptions you are facing.

From discussions with hundreds of managers from different industries over the past five years, I have distilled four questions that you want to ask as your company faces the prospect of rapid digitization.

Your relevance

What will your role and relevance be when
your industry has transitioned to a digital world?

Digital technologies challenge the product architecture road map of nearly every industry. Sensors and always-on cloud connectivity change the way services get delivered. As a result, digitization may alter the role your business plays in the industry and your ability to earn revenue and realize profits. Once you understand this, you can assess your pockets of value and arrive at a business model where your role is differentiated and relevant. Your offering may remain the same, but you may need to bundle it in creative ways to deliver compelling, customized value to consumers. Or the transformation could be more profound; if you're a product company, you may need to become a service company (and vice versa), or you may need to create platforms for products and services from others. If you're a service company, you might need to deepen your customer relationships to become a solutions company. In Chapters 6 and 7, I'll provide a systematic way for you to understand the different options and interconnect with others to ensure continued relevance.

Your distinctive capabilities

What distinctive capabilities does your company have that
will allow you to maintain your relevance in a digital world?

Digital technologies replace or modify some traditional resources and sideline others. For example, algorithms, apps, and bots can automate processes and replace significant segments of activities in industries. What distinctive capabilities does your company now have that could still be relevant in the digital age? Or how should you modify your capabilities so that they are relevant for the future? In order to possess those distinctive, differentiated capabilities, you may need to think outside your historical core competency, and reallocate or rethink your traditional financial structure or hire people with different skills. How could you use cognitive computing or machine learning to enhance these capabilities? I'll discuss this subject in more depth in Chapters 7 and 8.

Your key relationships
How should you partner with other players?

Digital strategy is network-centric. That is, value is co-created in the digital world when businesses work with partners and customers, designed right from the beginning. So your success depends on how you build your network of relationships and change them dynamically over time. Again, these relationships are more than the ones you may have structured and managed to execute today's strategies. Remember, these are key relationships to help shape your strategy in the digital era. I'll look at this question in Chapters 6 and 7 and provide you with approaches to navigate effectively in dynamic ecosystems.

Your transformational path(s)
How should you effectively mobilize
and transform for the digital future?

Digital shifts call for faster and more dramatic changes than just embracing some of the latest digital technologies that work with your past organizational architecture. You may need to revise and speed up your product road maps; migrate your internal processes to the cloud; increase your codependency with software partners and networks; acquire and assimilate new capabilities; recruit employees with newer, more relevant skills; and possibly create separate organizational entities. As a business leader, you must pave the way for these possible transformational paths, or your vision will remain unimplemented. And you must develop the rules that are likely to help you transform and win. I'll discuss how to successfully adapt and transform in Chapters 9 and 10.

THE DIGITAL BUSINESS game that you have stepped up to play is undoubtedly risky but potentially rewarding; it's a game that you cannot avoid or sidestep. But nor is it a game that you wish you didn't have to play. You now know the new types of players and their skills. You also appreciate that working together will bring out the best approaches to deliver value to customers, which will benefit your company, your collaborators and competitors, and your customers. And although some pundits have prematurely declared many incumbents dead, you well know that you still have many opportunities to reinvent your business as your industry digitizes along the three phases. Part 2 will help you do just that.

PART 2

THREE PHASES OF TRANSFORMATION

CHAPTER 3

EXPERIMENTATION
AT THE EDGE

{**ex • per • i • ment • a • tion** | the systematic act, process, or practice of learning}

WHAT COULD GE learn from Facebook? Kyle Reissner was a member of the product management team when he asked himself this question. "Facebook's genius is how the back-end systems dynamically adapt to real-time situations with historical intelligence through simple to use interfaces. That is how they are elevated beyond just a social network, to become a platform for real-time, social intelligence," he thought. And he wondered what might happen to GE's efficiency if it "swapped the users out as data sources and put machines in?"[1] In other words, where Facebook makes use of data from humans to push the right information to the right person at the right time, GE could do the same with information from sensors, machines, and systems. This line of thinking has helped GE to think about itself as "Facebook for industry" and develop experiments to explore real-time operational intelligence solutions. Looking beyond the obvious, connecting the dots, and developing a compelling narrative to guide your future business decisions is a valuable reason to experiment and begin phase I of your company's digital transformation.

SCAN WIDELY BUT CONNECT THE DOTS

If you think back to the nine screens on your Digital Matrix control panel, the three screens in the first column represent the various experiments carried out by the three sets of players. These are the different initiatives from maverick entrepreneurs, bold experiments from digital giants, and thoughtful extensions from incumbent companies in different industries. They're experiments unconstrained by past definitions of industry or functions or geography. You want to scan widely, but your objective is clear: to make sense of the experimentation frontier and guide your company to transform. Your ability to connect the dots across a wide range of experiments by the three sets of players is key to how your business wins in the digital future.

You are acutely aware that your current business models are based on rulebooks that have helped your business to win in the past. But your role as a leader is to reassess the assumptions underlying those rulebooks—the ones that your organization currently follows—and examine how and where they should be revised. Making sense of experiments helps you with this revision.

You know that future business models lie at the nexus of seemingly disparate, diffused, and disconnected trends such as the social web, the blockchain, artificial intelligence, and wearable technology. And you see where these disparate trends might converge and create compelling new business rules. This is more than coming up with mobile apps to port web content to smartphones and tablets; it's more than creating me-too Facebook pages for your brands or striving to imitate Uber or Airbnb in different industries. You are especially looking to discover those cases where ideas that might have been dismissed as science fiction one day, like 3D-printed cars or human organs, have become very real and commercial successes the very next day. Understanding *how, where,* and *when* they could become real in prototypes and at scale is important.

You are scanning widely to develop a story—a compelling business narrative well supported by data and analysis—of how digitization could shape the future of your industry. You are curious to see how disjointed technologies could interconnect to unleash new business functionality, just as mobile and social webs intersected over the last five years. You are interested in knowing precisely how distinct technologies such as 3D printing, robotics, and drones emerge as driving forces to redefine the geographical distribution of high-value manufacturing activities. Your challenge is no less than developing a theory of digital business by laying out plausible scenarios from experiments occurring at the edge of your own industry and in other industries, too.

To use a photography analogy, you need a powerful lens that combines a wide aperture and a long depth of field. Interestingly, all three sets of players may be using the same type of lens but looking for different objects and trends. That's what makes this phase challenging and powerful. That's why your own approach to make sense of the experiments at the edge and develop implications for incumbent leaders is important.

ENTREPRENEURS EXPERIMENT TO LAUNCH AND EVOLVE THEIR BUSINESS

In 2010, UberCab (now Uber) launched a mobile app in San Francisco that has since changed the taxi and transportation industry not just in that city but globally. The functionality of the app itself is rather straightforward: with a tap on a smartphone, Uber links people who need a ride with drivers who can provide one, at a price driven by market demand and customized to the rider's needs at that time. Clearly, no other app could make that claim, at least until now. Uber's global prominence in five years reflects what digitization could unleash when unfulfilled market needs are met by pulling together different technologies at scale and speed. Let's take Uber as an experiment at the edge of the traditional automobile industry and ask three questions:

1. Should global car companies such as GM, Ford, Toyota, and Mercedes-Benz have experimented with such a ride-sharing service early on? One argument suggests that if they had seen their role as providers of transportation service and not just designers and builders of

automobiles, they could at least have experimented by extending their smartphone apps beyond telematics (car unlock, navigation, and monitoring engine performance, as in the case of GM OnStar). GM experimented on a limited basis early on with RelayRides (which has since been rebranded as Turo to be the Airbnb of car sharing) but not at the scale of what Uber has now become. The question is rhetorical because we know the answers, but the line of thinking is useful, as it raises interesting and important challenges as we try to understand this early phase of digital transformation. If incumbents, such as automakers, can recognize the potential implications of such experiments at the edge of their business model, they will have time to figure out the best set of responses.

2. Should car rental companies such as Hertz and
Avis Budget Group have experimented with this model?
Avis Budget Group did acquire Zipcar—arguably the first car-sharing service (albeit self-driven, not driver-supported) for about $500 million in 2013,[2] just as Uber was expanding globally. One could make the case that Uber needs GM, Ford, and the like to produce the cars to make the Uber model work. Similarly, we could argue that GM needs Uber, just as it needed rental car companies as regular captive buyers. But what about Avis Budget and Hertz—does Uber threaten the rental car business? Did rental car companies perceive Uber as complementing their business models but not threatening them?

3. What does the Uber experiment mean for incumbents
in industries far removed from automobiles and rental cars?
Could this model apply to boats or planes or recreational vehicles and other areas beyond transportation, such as tailors, plumbers, dog walkers, and electricians?[3] By using the phrase "experimentation at the edge of your business model," I am suggesting strongly that experiments such as Uber have implications far bigger and broader than the immediate industry in which they start or the specific problems that they are first designed to solve.

As you sit at the control panel looking at the three screens in the first column, you search far and wide to see what other areas are ripe for Uber-type experimentation.

Learning from Netflix

The story of Netflix, which disrupted the movie rental business and perhaps caused the bankruptcy of Blockbuster and other video rental chains, may be well known to many. But the more important chapter of the Netflix story involves the experimentation underway right now. The DVD-by-mail business, which was the core of Netflix when it started, is a tiny part of the company today. Through video streaming, the company has steadily transformed itself into a leading Internet television network that could replace the broadcast TV business model of the last century. Internet TV is on demand (time shifted, place shifted), personalized (unique based on individual viewing habits and predictions based on many data elements including Netflix links to Facebook), delivered from the cloud to any screen (device shifted) anywhere (subject to legacy legal restrictions). More than 75 million subscribers pay their fees in more than 190 countries, and, each day, these subscribers enjoy 125 million–plus hours of TV shows and movies.

Reed Hastings, the co-founder of Netflix, has been a consummate experimenter. Multiple trials over time have allowed him to steadily fine-tune his business model for the media web of the twenty-first century. All of his experiments have been at the edge of the television and entertainment industry, but the mainstream industry leaders have never taken them seriously, until perhaps now.

While most others in the movie rental business were focused on top movie hits based on general lists, Hastings focused on developing a "recommendation engine," software that would learn an individual's film preferences and use that information to suggest other appropriate titles from its catalog. At first, the engine relied on individual ratings and preferences from previously watched films to predict user ratings for other movies. Later, it pulled together data from similar profiles of viewing habits. To further ensure the superiority of the recommendation engine, Hastings introduced the Netflix Prize, an open competition in 2006 for anyone to beat the company's Cinematch filtering and recommendation algorithm. He wanted the best technical minds to challenge the algorithm so that it would continue to be at the leading edge.[4]

Then he experimented to add unlimited streaming (albeit with a very limited content library during the early years) to his DVD subscriptions in 2007, which allowed him to understand early on how people were embracing online streaming, despite all the limitations of bandwidth and unreliable cellular networks.

Today, Netflix is the undisputed leader in video streaming, and personalization is a big reason for that success. Each subscriber's homepage shows groups of videos arranged in rows with a title that conveys the meaningful connection between the videos in that group. Netflix uses its filtering and recommendation algorithm to select and personalize the rows, the items for each row, and the order of the videos in each row for every individual. Netflix is a leader at personalization, which is done by machines and at scale and speed.

Hastings' experiment to work with Amazon for cloud-based streaming (instead of owning this digital functionality) was a well-reasoned analytical decision: there was no way Netflix could compete against Amazon's cloud competence.[5] (I will discuss the importance of working with competitors in Chapters 6 and 7.)

By systematically analyzing the pros and cons of creating Netflix hardware (like a Roku or Google Chrome or Apple TV) versus focusing on software, the company ultimately settled on Netflix software to make streaming work across multiple devices—television screens, set-top boxes, Netflix-ready television sets, personal computers, tablets, phones, and others.

Whereas other television networks compelled executive producer Kevin Spacey to prepare pilot shows to test audience reaction to his proposed show, *House of Cards,* Hastings and his team were able to assess its prospects by analyzing the treasure trove of data about subscriber viewing habits.[6] And whereas networks traditionally select shows based on a pilot, followed by a few shows, then an entire season, Netflix, using its data to mitigate any risk, was able to commit to two seasons up front—which allowed the producers and directors to be more creative.

Based on the success with this show, Netflix has gone on to create other original programs such as *Orange Is the New Black, Bloodline, and Marco Polo.*

So, at the edge of the media and entertainment industry, Netflix had been experimenting seriously and serially with different building blocks for Internet TV. Most traditional incumbents, such as Time Warner, Disney (ABC), Fox, NBCUniversal, and CBS, woke up late to the power of recommendation software that drives the personalization that has become the hallmark of Netflix. These companies were also late to recognize the power of data-driven insights that allowed Netflix to get into original programming, and most did not understand Netflix's technological superiority to optimize streaming for different devices and across different bandwidths from distributed cloud architecture. Such sequenced experiments, refined over time, allowed Netflix to become a global powerhouse in Internet TV. Television incumbents turned up late with their own ways to compete against Netflix; their versions of Netflix-killers through apps and alternative alliances.

As you make sense of Netflix, you realize that the experiments undertaken by Hastings and his team were not one-offs; they were sequenced to marry developments in technologies with plausible options for superior customer value and different business models that place digital characteristics such as algorithms and analytics as well as streaming and cloud as key anchors. As you try to make sense of the different experiments in the media and entertainment space, you may be wondering ahead to phase 3: Will the digital future of media and entertainment be defined by tech entrepreneurs like Reed Hastings and digital giants such as Amazon (Prime Video), Alphabet (YouTube), and Apple (iTunes and Apple TV)? That question about how the foundations established in the experimentation phase could drive the reinvention phase is not restricted to Netflix. You will find that linkage to be an essential thread of the Digital Matrix.

Uber looked like a minor smartphone app for hailing taxi service in 2010.[7] Netflix looked like a DVD-by-mail distribution company in 1997, started by someone who hated paying late fees! These two have become strong contenders in traditional industries that have progressively digitized. Their experiments, individually, may not have looked significant to the incumbents in those two industries—but, over time, these experiments have allowed them to master digital capabilities that form the basis of powerful, profitable business models. These and other entrepreneurial experiments that appear at the periphery often look preposterous and far-fetched or minor and inconsequential. But they can quickly gain momentum, morph with developments in technology, seize on new features, and, with partners, evolve an ecosystem and grow rapidly.

INCUMBENTS LOOK AT EXPERIMENTS IN TWO WAYS

When managers in incumbent companies ask me, "What digital experiments should I track?" my answer is simple. You should look at two types of experiments at the edge as early signs: those in which experimentation *complements* your current business models and those in which experimentation *challenges* your current business models. The former gives you early indications of how you might proactively embrace those ideas, and the latter points to warning signs to consider. So as you now sit to look at patterns of actions and interactions within incumbent companies, you should adopt these two points of view.

Experiments that complement: Learning from Nike

In 1987, Nike introduced a device called Monitor, which was about the size of a paperback book. It had a strap with sonar detectors that tied around a runner's waist and could calculate a runner's speed and announce it through the runner's headphones. This was the time of Sony's Walkman, well before the iPod craze, and Nike was very much rooted in the business of sports footwear and apparel. Nike had always dabbled in digital technology to a certain degree, and this was one of those experiments. The Monitor failed; it was clearly way ahead of its time.[8]

Nike then tried selling branded sports watches and heart rate monitors, and they failed; again, they were ahead of their time. If the company had not embraced experimentation as a way to understand digital futures, Nike's senior managers would have killed any further initiatives with new technology. But Nike persevered and persisted in its belief that digital features could complement their physical products, experimenting with designs that led to the release of a smart shoe—one embedded with sensors—in 2004. This was way before the emergence of a network of smart devices that we now call the Internet of Things (IoT); so the smart shoe was, again, well ahead of its time.

What made this experimentation distinctive and instructive was that it marked a joint collaboration between two companies—one in the traditional industry (Nike, whose expertise was in shoes and apparel) and another in the nascent digital sector (Apple, whose expertise was in sensors). Phil Knight, the founder of Nike, saw how primitive sensors—despite their limited functionality—could be used in shoes to provide runners with feedback so that they could modify their behavior and use. Nike worked with Apple to create Nike+, a running shoe that logically connected running (its core competency) with music and data collection (complements). Or as Mark Parker, the current CEO of Nike, remarked about the Nike+ experiment: "Most runners were running with music already, and we thought that the real opportunity would come if we could combine music with data."[9]

Together, Nike and Apple explored the frontier of joint experimentation at the edge during a period of nascent digitization. Apple had released its portable music device, the iPod, in 2001, and the iTunes technology allowed Nike to seamlessly sync data at that time through a personal computer acting as the interface. Apple refined Nike's prototype sensor while Nike focused on the shoes and the interface for the Web and the iPod. The result was what Apple's CEO Steve Jobs called the "speedometer for sports." In other words, a runner wearing a shoe with Nike+ was able to stream music and collect data about her run. Essentially, it was the beginning of the "quantification of the self" movement, in which we now use apps to measure how we live, work, and play. This kind of collaborative arrangement between traditional companies and digital giants is an essential characteristic of the Digital Matrix.

What can your company learn from this example? You need to recognize the signals at the periphery that could prove interesting and potentially important in the future. In 2016, the idea of being able to obtain a custom-fit running shoe at your local athletic footwear store may appear far-fetched. Yet companies such as Nike, New Balance, and Adidas should be (and probably are) experimenting with 3D printers to make this idea a reality. As Mark Parker remarked: "If innovation drives everything we do, then digital is an accelerator. Digital has fundamentally transformed our business end to end, impacting really how we design, how we manufacture, and how we serve the consumer. It expands creativity and creates scale."[10] How far could such an idea go? Nike could extend its business model to take over manufacturing, becoming the owner of small local retail locations that resemble high-tech design studios with 3D printers that produce on-demand shoes for individual runners,[11] customized to their foot and embedded with sensors that track their individual fitness. To get there, Nike would have to rethink its whole system of business activities, from design to production to distribution and use—and develop a new series of experiments to test the market.

Experiments that complement:
Learning from Under Armour

One of Nike's fiercest competitors is Under Armour, so it's worth looking at how that company is tapping into technology to carve out its own niche within the highly competitive sportswear industry. In 1996, Kevin Plank, a university football player, began experimenting with a moisture-wicking synthetic fabric that he made into a T-shirt to keep him from drowning in sweat during games. He passed around his prototype to teammates and friends who had gone on to play in the NFL, and word of these shirts began to spread.

By 2013, Under Armour was using e-commerce to take orders for its products and had developed a heart rate monitoring strap that attached to its shirts. But Plank understood quickly that the landscape of digitization was broader and deeper, and that his company could become obsolete if he did not comprehensively embrace and exploit digitization. So he acquired MapMyFitness, Endomondo, and MyFitnessPal, three fitness-training apps, and gave them to three hundred engineers and app developers.[12] Taking a page from Facebook and the social web, in just three years, Under Armour had created a connected fitness community with more than 150 million active members.

How did Under Armour do this?[13] First it experimented with collecting data from different types of sensors embedded in its shoes and shirts using technology from a partner, Zephyr Technology. Then it analyzed the real-time information from these sensors—e.g., heart rates, skin temperature, and acceleration speeds of athletes using the products. Beyond selling products, it branched out to create vibrant communities such as LinkedIn and PatientsLikeMe. Under Armour has been collecting the data community members have been actively uploading about their lifestyles, such as the food they eat and the gear they use for fitness. And like Netflix and Amazon, which have redefined the state of recommendation engines with big data and analytics, Plank sees this customer data as the fuel for redefining his company for the future. It has to develop a deeply trusted relationship, "to know everything in your locker, every time you run, everything you put in your body, all the stuff you need, all the content you need, all your likes, all your dislikes, and to build that into one relationship where we can really solve problems for you." For Under Armour to succeed, it needs to earn the trust of its members and bring in a wide range of partners, such as Walgreens and Humana. It must continue to keep innovating to ensure that its platform is *the* place that individuals go to talk about their health and fitness. Kevin Plank looks at Amazon as a reference point: "Forty percent of Amazon's revenue comes from their recommendation engine."[14] He wants Under Armour to make recommendations on behavior, using the data that the company's online community has entrusted it with.

In January 2016, Under Armour announced an important relationship with IBM to take advantage of that company's cognitive computing functionality (Watson) to develop data analytics and provide personalized services to customers.[15] These efforts reinforce both their brands, but the information about how Under Armour's consumers use its products enables the company to develop deeper insights about customer engagement, which in turn allows for faster and more effective feedback from experiments and perhaps greater opportunities at personalization and customization. In the digital world, companies that have customer feedback about their products in use under different conditions have a much better chance of success than companies that have only research data about their products in ideal conditions in a lab.

In both Nike and Under Armour's cases, their experiments complemented their existing business models, allowing them to test ideas to refine and adjust their product offering. These experiments also showed how the leadership—starting with the CEO—understood the importance of digital technologies. Both CEOs were personally involved in guiding their company's digital experiments and were prepared to work with external partners and make necessary acquisitions to get started on their digital journeys. They are not alone. Scan your own industry and others to understand how incumbent companies in different industries have been working through ways of accepting and absorbing digital functionality. Such assessments can help you develop deeper, more focused insights about how experiments at the edge of traditional business models are being pursued. What you'll find is that many companies are experimenting with ideas that complement their existing business model; however, some companies have embarked on experiments that could actually cannibalize their historically successful model. Let's look at this situation in which companies are working hard to overcome their success traps.

Experiments that challenge: Learning from Ford

Leaders in some companies are prepared to accept the possibility that their current business models—and their accompanying assumptions about impermeable boundaries among industries, narrow scope, functional specialization, value chains, etc.—are out of date for the digital age. When that happens, they're prepared to experiment not just with refining or adjusting their tactical plans but also with re-examining their approaches completely. These incumbents have woken up to the fact that such experiments are necessary. If they fail to examine potential disruptions and continue to stick with incremental evolution of the status quo, they stand to run into formidable challenges from upstart entrepreneurs and digital giants. Their very existence may be under siege. That said, I have observed much more willingness on the part of incumbent leaders to experiment with digital technologies in recent years than ever before, particularly in companies in which the top leadership and senior management have understood and accepted the power and pervasiveness of digital technologies. They know well that the pivot from the industrial-age model is immediate and that they cannot simply hold on to their historical theories and assumptions.

Mark Fields, the president and CEO of Ford Motor Company, has undertaken more than twenty-five different strategic experiments to examine the future of transportation at the nexus of traditional industrial and digital technologies.[16] In his words: "We have given our engineers, scientists and technologists a challenge. We have asked them to use innovation not to just create better products. We have asked them to innovate to make the entire transportation experience easier, to make people's lives better and, in doing so, to create a better world."[17] These experiments examine areas as diverse as remote-controlled cars that use 4G networks, smartphone apps to help find open parking spots, and systems to control a smart home's thermostat from car dashboards. Individually those experiments might seem trivial and incremental. But collectively they could lead Ford away from designing and producing automobiles to delivering multimodal transportation services not necessarily anchored to Ford cars or trucks.

Fields describes these experiments as "evolving our thinking, adjusting our view of the car as an individual object to the car as part of a broader transportation network."[18] In March 2016, the company established a separate subsidiary—Ford Smart Mobility—with operations in Palo Alto, California, and Dearborn, Michigan, to expand the business model as an auto and mobility company.[19] That shift away from a belief that cars must be individually owned is a far cry for a company that ushered in the idea of personal car ownership as individual freedom in the twentieth century.

These experiments seek to answer the four key questions to guide strategic thinking that I presented in Chapter 2: they help Ford articulate its vision for the future as a digital business, help identify and develop its distinctive capabilities internally and externally with partners, and clarify which resources must be mobilized to move the organization forward. Bill Ford, the company's chairman, believes that although some of these experiments will fail, they will offer the necessary insights to help his company compete and morph into a different business than his ancestors created in the early twentieth century.[20] Could Ford go the way of IBM and rely less on hardware and more on software and services? I will show in Chapter 4 how different companies are colliding at the core of the automotive sector.

Experiments that challenge: Learning from GE

Just like the Ford Motor Company, GE has been experimenting widely and with reinvention in mind. Over the last three decades, GE's success has resulted from being a conglomerate that accessed and distributed its financial capital across a widely divergent portfolio of businesses to deliver superior returns to its shareholders. GE, under the leadership of Jack Welch, was in a class by itself. It succeeded wildly while many other conglomerate organizations were breaking apart. Under its new CEO, Jeffrey Immelt, GE is reinventing itself as a "digital industrial company." He has been leading GE's experiments at the intersection of physics and data—where its historical competency with materials science is now coupled with data sciences. In Immelt's words: "I have never made a mistake in my conference room; only the market determines success. Winning in the market requires experimentation."[21] He believes that the company's Industrial Internet initiative has been its most important experiment.

Immelt explains, "There are some businesses that design software, and others that focus on building and manufacturing. But GE is the only company that can join innovative technology with industrial depth."[22] By this he means developing an industrial Internet of Things, products connected to smart machines that collect data and use powerful software to analyze, optimize, and customize the products' usage for individual customers. He has been motivated to learn from the first wave of digitization within consumer industries, asking whether asset-light, data-rich industries mostly focused on consumers with advertising and media could have any parallel applicability to the asset-rich, data-rich industries that GE has operated in. Specifically, he is interested in the potential of new value that could be created with digitization. "If you think about today, 15 percent or 20 percent of the s&p 500 valuation is consumer Internet stocks that didn't exist 15 or 20 years ago. The consumer companies got none of that. When you look at retailers, banks, consumer-product companies, they got none of that. If you look out 10 or 15 years and say that same value is going to be created in the industrial Internet, do you, as an industrial company, want to sit there and say, 'I don't want any of that. I'm going to let a Newco or some other company get all that'? Is that really what you've relegated yourself to?"[23] In other words, without an aggressive digitization agenda, GE, or any other industrial incumbent, for that matter, will find its value commoditized very fast in the future.

Mark Fields at Ford and Jeffrey Immelt at GE may be two of the more well-known senior leaders spearheading headline-worthy experiments to reinvent their organizations for the digital future, but they are part of a small but growing cadre of leaders that have realized the future is not an extrapolation of the present. They know that past success does not guarantee future success and growth, and that reinvention is necessary, now.

In general, tech entrepreneurs have a specific vertical focus targeted at a particular industry and problem—sometimes experimenting to help industry incumbents and sometimes challenging the traditional ways. Incumbents are looking to complement and extend their traditional strengths in an industry or discard an outmoded business model and reinvent themselves for the future. Digital giants approach experimentation at the edge differently. As I discussed in Chapter 1, they have scale and scope ambitions to differentiate themselves from their direct competitors, other digital giants. So their experiments typically involve how they could achieve increased scale by newer means or how they could expand their scope into new domains. Let's look at some of the exemplary experiments pursued by the digital giants in areas outside the traditional information technology sector.

DIGITAL GIANTS EXPERIMENT TO EXPAND THEIR INFLUENCE ACROSS INDUSTRIES

In Chapter 1, I discussed how digital giants have steadily expanded their scope of influence and effect on industries that are distant from their core domain. Just as tech entrepreneurs pitch to venture capitalists every day about how their experiments will change the business world in different sectors, digital giants also explore how they could extend their influence into different industries. Let's look at two different arenas—health care and conversational commerce—to sketch out the key ideas.

Learning from health care

Health care is increasingly interesting as a field in which digital giants are innovating and broadly expanding their scope. In fact, digital health experiments have been underway for many years, but nearly every digital giant now has at least one health-related project in an experimental stage. Google might have abandoned its previous incarnation of health—a place to store consumers' health data—in 2011,[24] but the company has not abandoned its focus on this sector. Under its new governance structure, Alphabet has dedicated an entire unit,

Verily Life Sciences, to exploring the question: "How can we use technology to create a true picture of human health?"[25] It has established multidisciplinary teams comprised of chemists, engineers, doctors, and behavioral scientists to work together to understand health and ways to prevent disease. That's not all, as Alphabet has joined with a Johnson & Johnson unit to create Verb Surgical, a company designed to create "the future of surgery at the nexus of machine learning, robotic surgery, instrumentation, advanced visualization, and data analytics."[26] In other words, it's a high-tech company to redefine surgery as it's practiced today. And Alphabet, with its focus on health, life sciences, and longevity,[27] will be a force to be reckoned with in the coming decade.

Apple has taken a different approach to health care.[28] Its Health-Kit is smartphone software that allows developers to create a wide range of health and wellness apps with sensors built into the iPhone and Apple Watch. Its ResearchKit develops apps that allow researchers to collect robust medical-grade data for research purposes, and its CareKit develops apps so that individuals can take a more proactive role in their fitness and well-being. Indeed, Apple's health-related software frameworks have sparked extended experiments in health care labs and medical research institutions. Time will tell how far Apple's software kits influence the future of healthy living. Microsoft's Health-Vault experiment[29] gives individuals a centralized locale to gather, store, use, and share health information for the entire family. And, of course, IBM has been using its cognitive computing technology, Watson, in various health care settings[30] that I will explore in more detail in Chapter 4. Although experiments by the digital giants show bold ambitions that they could bring their competence in algorithms and analytics to transform health care, many industry incumbents are skeptical. They wonder how the giants will deal with digitizing health records, which are currently scattered at the level of the hospital and patients and involve issues of privacy and security. It seems inevitable, however, that this sector will be transformed as patients demand better quality of care at a lower cost than they are getting today. If you are an incumbent in the health care sector, take note. The digital giants are achieving horizontal integration across your industry, as they have done in several others already.

Learning from conversational commerce

One of the fastest-growing areas of experimentation in 2016 is conversational commerce, which means using "chat, messaging, or other natural language interfaces (i.e., voice) to interact with people, brands, or services and bots... The net result is that you and I will be talking to brands and companies over Facebook Messenger, WhatsApp, Telegram, Slack, and elsewhere before year's end, and will find it *normal*."[31] Many companies, such as the Hyatt hotel group, have been experimenting with messaging apps (e.g., Messenger) for direct conversations with consumers, and they've been getting great feedback from their clients.

Next are customer service bots. In contrast to today's digital personal assistants, such as Apple's Siri or Amazon's Alexa, which provide answers to direct questions, these new chatbots can carry on a conversation. The difference is measurable. Analysts count the number of turns in a conversation, essentially the number of times speakers alternate, and record this number as conversations per session (CPS). Today, a typical interaction between a human and a digital personal assistant has a CPS of 2: the human speaks once and the chatbot speaks once. That's a question-and-answer session. However, Microsoft has developed an experimental chatbot for the Chinese market named Xiaoice (pronounced Saho-ice),[32] which averages 23 CPS with its 40 million users.

What's significant about this result? Two things. First, Microsoft can now begin to experiment with "emotional computing," which means conversations that require a balance between analytical and emotional intelligence. That's the kind of interaction typically associated with humans, especially with doctors and their patients or with teachers and their students. Second, "through the tens of billions of conversations she's had over the past eighteen months, Xiaoice has added considerably to her store of known conversational scenarios and improved her ability to rank answer candidates. Today, 26 percent of the data in Xiaoice's core chat software derives from her own conversations with humans, and 51 percent of common human conversations are covered by her known scenarios. *We can now claim that Xiaoice has entered a self-learning and self-growing loop. She is only going to get better* (emphasis added)."[33]

Not to be outdone, Amazon established a $100 million venture fund to pay developers to create apps (or, as Amazon calls them, "skills") that could be embedded into products, including its Echo and Echo Dot wireless speakers, to encourage conversation.[34] For example, you can expand the scope of Amazon's Echo speakers to play, say, the *Jeopardy!* trivia game by adding a skill from a developer. Or you can add a different skill and ask Alexa to summon a car from Lyft or adjust your Honeywell thermostat.[35] This conversational platform could be used to order (and easily reorder) items from Amazon, but more importantly, it positions the company to compete against Apple and Alphabet in the emerging architecture of smart homes. Like the operating systems in personal computers and smartphones, this digital technology allows the company that owns it to control the transactions that occur on or around it. As you think about your own experiments, consider how conversational commerce might change the balance of power between you and your customers. The digital giants are beginning to achieve vertical integration, which means that as your industry further digitizes, they will exert an ever-increasing influence on your business.

With so many players experimenting in such a range of different ways, how do you assess what's relevant for you and your business? As best you can, look at each of the players in turn and pick out the ones that directly *challenge* or *complement* established practices in your industries and markets. Analyze patterns from their patent applications. Look at the research findings presented by their scientists at high-profile conferences. Formulate your own rough narrative. That narrative sets the backdrop for your response.

TWO LEVELS OF RESPONSE:
FROM OBSERVE TO COMMIT

Looking at experiments at the edge of your industry and others from the perspective of digital giants, tech entrepreneurs, and your other industry incumbents gives you a better sense not only of the wide-ranging ways in which companies are testing their business models and the market but also of the strategies that are motivating them. The leaders of all these companies share one thing: a curiosity to explore the frontiers of digitization where none of the previous ideas of product definitions, industry boundaries, or value propositions matter. Leaders of successful digital companies see fascinating possibilities and carry out experiments—often with others—to develop promising pathways to create their digital futures. Remember, these are not technical experiments but business ones; they are not single, isolated initiatives but sequenced and coordinated as a portfolio of interdepen-dent ones.

With these broad ideas in mind, what steps should you take to think about experimentation at the edge of your industry? What theories of future digital business could you understand in more depth with a series of focused experiments? The responses of incumbents to experiments at the edge fall into two levels of increasing importance and acceptance. The first level of response is observation and the second level of response is commitment (or investment).

OBSERVE

If you and your management team have recognized and accepted the possibility that digital technologies could influence and change the course of your industry, but you are not yet convinced that digitization has become mainstream and core relative to other priorities, you're most likely to *observe* what's happening. Think about this level of response as fine-tuning your scanning antennae to sense and track signals that could indicate the shape of things to come in the near future. How can you do this? Immerse yourself in the flow of digital innovations covering a wide terrain.

When I ask executives to list the digital business experiments that they track at the edge of their business models, I get mostly reactive responses. They track what their competitors do with digital technologies, including social media experiments with Facebook, Google, or Twitter; new enterprise mobile apps and consumer apps; IT and digital technology spending, new alliances, and partnerships; and so on. In other words, they mostly follow fairly standard competitive scanning and analysis protocols. And they get this information through a service that provides filtered business headlines on specific themes (competitive intelligence analysis or investment alerts) or through consulting company reports.

Since digitization is a multi-faceted force, I've found that a systematic approach is helpful. The following questions are a useful starting point to *observe* experiments at the edge.

1. What are your current competitors doing with digitization internally? What experiments and announcements from digital giants and new entrepreneurs could potentially have relevance for you?
2. What joint experiments have your competitors carried out with the two other sets of players? How significant could they be in shaping the future of your industry?
3. Looking beyond your industry boundaries into other industries, what experiments should you continue to track to develop proactive insights?

The first two questions are straightforward, but the third is more significant, as digitization ideas for innovation, disruption, and transformation do not lie neatly within predefined industry boundaries or in industries that have historically looked similar to yours. For example, GE benchmarked Facebook to develop insights into how it could architect its real-time operations dashboard; it also benchmarked Apple and Google as leaders in consumer mobile operating systems and software applications to assess how it should think about platforms and apps for the industrial world. In the same vein, you should examine several promising areas. For example, how could Amazon's experiments with conversational engines (i.e., Alexa) or recommendation engines help you if you are in the financial services or residential energy or industrial transportation industries?

Boardroom conversations in retail companies today routinely deal with challenges from digital upstarts that many label as intruders, but every company should explore how automation algorithms and robots could change the way work is designed and carried out not only in high-tech sectors but in every sector. As an incumbent, you may be willing to close your eyes to the possibilities of digital disruption, but the digital giants and entrepreneurs are constantly looking at creative and compelling ways to take advantage of digital trends to disrupt the status quo. And you should be too.

Establish a "sense-making unit" in centers of digital innovation. If possible, locate a few (never just one) of your employees in Palo Alto (or, for that matter, in Tel Aviv or Bangalore or Shanghai or New York or London or Berlin or Boston) and have them look for critical trends that could both enhance your current business model or challenge it. They are your "sense makers"; these executives can minimize your risk of not knowing where and when new disruptions from technology could hit. They mingle with entrepreneurs; host and participate in meet-ups; attend open sessions and conferences at universities and think tanks; and organize, participate, and judge hack-a-thons and open challenges. The sole mandate of these employees is to observe signals, prioritize the ones that may be of interest, and develop a narrative in the language and vocabulary that your organization understands. First, they observe how other incumbents are using their outposts to make sense of the different experiments. Second, they make note of what the digital giants are up to. And third, they report on what types of tech entrepreneurial ideas are gaining traction with the venture capital community.

If you are a telecom operator, you're interested to see what others are doing in innovation centers such as Silicon Valley. Swisscom, the

Swiss telecom operator, uses its Silicon Valley outpost to interact with entrepreneurs to understand the depth and breadth of mobile apps that could be offered to its enterprise and individual customers.[36] BBVA, the Spanish bank, initially operated an outpost in San Francisco that identified long-term shifts and decided on minority equity investments. To gain broader access to the deals in the financial tech sector, however, it closed its own office and joined Propel Venture Partners in 2016.[37] This was a strategic decision, based on observation and experimentation, not to make high levels of returns on investments but to be in the flow of venture deals and decide on the subset of trends that could be relevant to the bank.

If you are a product manufacturer, your sense-making unit might be a digital innovation lab. Levi's, like Lowe's and Home Depot, has set up its Eureka Innovation Lab[38] to experiment in different areas that have an influence on its business. Some of these experiments are solo projects, whereas others are collaborations, such as Google's Project Jacquard. In simple terms, "Project Jacquard makes it possible to weave, touch, and gesture interactivity into any textile using standard, industrial looms. Everyday objects such as clothes and furniture can be transformed into interactive surfaces."[39] In this experiment in creating wearable technology, Google is working with Levi's to create clothing with embedded electronics and interactive yarns. The idea is to make it possible to answer your phone or bring up Google Maps with a touch to your sleeve. It is likely to influence not only the digitization of the fashion industry but also other industries further afield.

Sense-making units also try to discern and develop a compelling narrative about the pattern of venture funding. If you are in health care, for example, you may be trying to make sense of the data from Rock Health, a Silicon Valley fund focused on digital health. If you are in energy, you may be observing the kinds of venture investments made by Braemar Energy Ventures in New York and London, or Vantage-Point Capital Partners or Khosla Ventures in Silicon Valley. Here, focus on the quantitative data for specific startups, but be sure to study the patterns of investment across the different venture funds to discern early trends.

Establish these outposts as the starting point, but realize that they are simply helping you collect and filter the relevant information for deeper discussions within your corporation.

Design workshops to probe deeper questions. Don't stop at just listing trends or creating blog posts or summary reports. Warren Berger, a noted author who writes about innovation, suggests designing "frame-storming" sessions: "When participants are generating questions, they tend to dig into a problem and challenge assumptions. For example, they may inquire about why the problem exists, why it's even considered a problem (maybe it really isn't one), whether there's a bigger problem behind that problem, and so on. The process gives people permission to ask fundamental questions that often don't get asked; not just how can we do it better but also why are we doing this in the first place?" The questions raised in such "frame-storming" sessions— brainstorming sessions focused on generating questions (instead of answers)—spark creativity.[40] With these questions in hand, observations become useful inputs and triggers for committing investments and follow-on actions.

Commit

If you are ready to commit resources—management time and financial resources—to those areas where digitization looks promising but uncertain, then you are ready for the second level of response. Think about this step as testing your ideas with data and analysis, using tactical experiments. Although the payoff could be big, and there will certainly be challenges, the priority here is learning. There is no failure, only deeper learning.

Design tactical experiments to examine the business impacts of technologies. Your focus at this phase is to understand what technologies to embrace, at what stage of maturity, to arrive at the level of invest-ment commitments, and the list of likely partners to create your advantage. In other words, this is about business experimentation with digital technologies. Ignore the technical feasibility of specific technologies—leave that testing to those vendors providing you with the technology—but look closely at the possibilities that digital tech-nology creates for your business, whether it be better data so that you can fine-tune your offering or connect with your customer, greater functionality so that you can broaden the scope of your product or ser-vice, or more efficiency so that you can become more nimble and pivot as required. More importantly, analyze to understand the potential threats and challenges from the set of technologies you have selected.

Siemens, the German global industrial company, has made the bold move of setting up a separate unit named next47 (a play on the fact that Siemens was founded in 1847) with about a billion euros over five years.[41] Its goal is to work with its own employees, external startups, and other industry incumbents to accelerate the application of disrup-tive technologies and position itself to compete against GE and other industrial giants. With its own offices in the US, Israel, China, and Ger-many, next47 is designed to be independent of the parent company but could tap into its expertise as needed.

Approach every experiment as a learning opportunity. The less you judge the likely results of your experiments—at least initially—the better. Instead of looking for a specific outcome, be open to the possibilities that the data suggest or how your consumers and partners respond.[42] Those results can lead you in new and unexpected directions that oth-ers aren't thinking about. Under the leadership of chief executive officer Bob McDonald, Procter & Gamble (P&G) carried out hundreds of dig-ital experiments with Facebook. Social media was new and held the promise of providing deep insights into individual customers, allowing P&G to fine-tune its marketing campaigns. Over the course of several years, P&G experimented with multiple iterations to develop rigor and analysis in the allocation of its marketing and advertising resources across different media, from print to TV, outdoor billboards, mobile, and social. More importantly, P&G gained valuable insights about the new types of skills it needed within its organization to best make use of the big data stream from Facebook and Google.[43]

What marketing experiments could you undertake with entrepreneurs and digital giants, using their algorithms and analytics to gain insights about your own customers? Are those experiments likely to give you more insights than the kinds of experiments your competitors may be pursuing right now?

Go beyond just passively observing digital trends and undertake your own tactical experiments to get your feet wet and test the waters. *Scan broadly* for experiments happening at the edge of every industry beyond your own, and keep your eyes and your mind open to how you could apply these trials in your own business. *Think laterally* about changes in the design and functionality of products. How could sensors and the data they generate transform your industry? *Be bold* by letting users play with and provide feedback on your experiments. Whether you share the concept, build a prototype, or partner with others, get your experiment out into the world. *Create community* by connecting with your audience or partnering with other firms that can help you achieve your vision. *Cultivate ecosystems.* Then weave those different tactical moves into your portfolio of coordinated experiments to prepare yourself for the next phase of digital transformation: collision at the core of business models. This is where digitization becomes real and intense, and your role to understand the implications becomes more profound and central than ever before.

CHAPTER 4

COLLISION AT THE CORE

{col • li • sion | a period of conflict between opposing perspectives or thoughts or forces}

WHAT COMES TO mind when you hear the names AT&T, BlackBerry, Blockbuster, Borders, Deere & Company, Dell, Hewlett-Packard, IBM, Intel, Kodak, Microsoft, Nokia, Oracle, Pfizer, Philips, Sharp, Sony, Texas Instruments, Toyota, and Xerox? Blockbuster and Borders may no longer be household names, but AT&T, IBM, and Microsoft are all still very much on businesspeople's radar, so it's not a list of companies buried by technological tsunamis. What all of these companies have in common is that they have faced challenges from newer business models put forward by tech entrepreneurs and digital giants. Blockbuster, Kodak, and Borders may not have survived, but the rest have so far successfully weathered this collision at the core of their business model and adapted. What can your company learn from their experiences?

ARE YOU PREPARED TO ADAPT AND TRANSFORM?

You may be inclined to think that some of the businesses that failed did so because they were information rich in products and services and asset light in business infrastructure, and that because your industry is asset heavy and information light, it is immune to the incursion of the digital giants and tech entrepreneurs restructuring your business model. I can tell you right now that it's a fundamental fallacy.

Sooner or later, incumbents in every industry are going to collide with tech entrepreneurs and digital giants—and maybe their own incumbent competitors—who have a different idea about how to design and deliver products and services, and how they can earn revenue and profits. Even if you and your industry are not yet in this phase, it makes sense to look and learn from those that have preceded you. Experiments at the edge can move very quickly—as many companies have discovered too late to ensure their own survival—and understanding how this happens can help you to be prepared.

If you think back to the nine screens on your Digital Matrix control panel, the three screens in the middle column represent the various collisions being experienced by the three sets of players.

SIGNS THAT A COLLISION IS TAKING PLACE

Have you started noticing companies from outside your industry beginning to challenge your business models? Have you seen innovations in value delivered to customers that take advantage of digital technologies? Have you seen experiments from yesterday evolve and become well-thought-out business ideas that challenge the traditional ones? If you find yourself in this phase, or if you believe that it is imminent, digitization should be on your management agenda *as a fundamental driver of growth and profitability for your company.*

Businesses typically start to take notice of the need for digital transformation when a collision occurs because it's hard to miss the competitive pressure that threatens your very existence. You will feel this pressure first in one of two areas, and then both.

Strategy collision. Your traditional strategic logic—and those of incumbents competing against you in the traditional industry—collides against newer logics that rely more heavily on digital functionality.

The newer models could come from other industry incumbents who have accelerated their digital transformation, from newer digital-born entrepreneurs, or from digital giants.

Organizational collision. Your organizational model—the traditional structure, processes, and systems born in the twentieth-century industrial age—collide against alternative ways of organizing born in the digital age. In doing so, they make your organizational model inefficient, and over time, this contributes to decline and extinction.

Historically, we separated strategy formation (which focused on business models and competitive strategies) from strategy implementation (which focused on organizational models and tactics for execution) under different hierarchical levels and functional departments. The core thesis of the Digital Matrix is that these two collisions are increasingly interdependent and occur together. Moreover, those companies born digital do not see any compelling reason to separate strategy and organizational models into functionally defined boundaries or hierarchies or even within and across corporate boundaries. To them, information technology is not a separate function under the chief information officer (CIO); it is the very essence of why they exist and how they plan to win. To them, there is no separate chief digital officer (CDO) or separate digital innovation unit. As Microsoft chief executive officer Satya Nadella realized when he took over Microsoft: "Any organizational structure you have today is irrelevant because no competition or innovation is going to respect those boundaries. Everything now is going to have to be much more compressed in terms of both cycle times and response times."[1]

In other words, even companies as large and successful as Microsoft are facing this collision at the core of their business models. Microsoft's past strategies and structure as a packaged software company will not withstand competitive threats from other digital giants and tech entrepreneurs whose business logic is defined by mobility and cloud computing. You, Microsoft, and every other company must reinvent themselves when faced with this collision. As an example, let's look at how this collision is occurring in one particular setting.

A home thermostat, redesigned for the digital age: Learning from Honeywell

A thermostat is designed to regulate temperature in residences and commercial buildings. The one offered by Honeywell is a typical industrial-era product, designed by Henry Dreyfuss in the US around 1956, long manufactured in the US, and sold through traditional channels such as Home Depot, Walmart, and Sears. Its design has changed very little in the past five decades. It has predictable sales, tied to the demand from the construction industry and the regular replacement pattern in residences, and its global sales and market shares matter. Its relationships with distributors such as Home Depot, Target, Lowe's, and Ace Hardware are critical, and its market position fluctuates only minimally. Honeywell leads the market in thermostats worldwide (it holds 30 percent of the market share), with Johnson Controls and Emerson Electric as other market-leading contenders.

In October 2011, a small startup company named Nest Labs introduced a new thermostat,[2] which it self-styled as a learning thermostat—it learns as users set and reset the temperature when they are at home and away, and after a week, it programs itself for energy optimization. The settings can be overridden as necessary through a smartphone mobile app.

If you'd been an executive at Honeywell, the leader in home and commercial thermostats, would you have been worried about this new product? Perhaps not, even if Tony Fadell (the designer behind Apple's legendary iPod) founded the company and declared, "It was unacceptable to me that the device that controls 10 percent of all energy consumed in the US hadn't kept up with advancements in technology and design."[3] You might have looked at this thermostat as an experiment at the edge of the industry with potential *future* impact but not yet worthy of more serious attention, even if the startup company was backed by the renowned Silicon Valley venture capital firm Kleiner Perkins. After all, venture capitalists routinely invest in technologies that are way off. Like many managers in industrial companies, you might have seen this thermostat as a new, specialized type of hardware and considered it less of a threat than a new software innovation or an e-commerce model.

In October 2013, the company introduced Nest Protect,[4] a smoke and carbon monoxide detector that uses Wi-Fi and cellular connectivity to alert your phone to a problem. If you'd been an executive at Honeywell, would this product extension at Nest Labs have raised concern? Perhaps a little, but you might have convinced yourself that this upstart company simply didn't have the size or scale to reach into all your complex distribution channels. You might instead have focused your attention on any competitive threats from Johnson Controls and Emerson Electric.

Three months later, in January 2014, Google acquired Nest Labs,[5] which had fewer than 300 employees at that time, for $3.2 billion. By November 2015, this unit had grown to about 1,100 employees, and all of them had a very different skill set than their counterparts at Honeywell or Johnson Controls. Nest Labs with Google's backing also acquired a webcam company and developed a platform for home security and automation. By early 2016, as an independent subsidiary of Alphabet, Nest had created its Works with Nest platform,[6] which links a set of its own products in your home via an app on your smartphone, and established an ecosystem involving major players such as Whirlpool, Fitbit, Mercedes-Benz, and others. Your Mercedes-Benz car could send the data on estimated time of arrival to the thermostat at your home to optimize the desired temperature in the house. Your Whirlpool washing machine could auto-delay its run cycle to take advantage of lower-cost energy during off-peak hours.

Now, if you were an executive at Honeywell, Johnson Controls, or Emerson Electric, would you be concerned?

You should be, because digitization of the thermostat (and smoke detector and other related products) is no longer a peripheral experiment. The core product's technology has shifted from analog to digital, and the core product itself is interconnected with others as part of broader platforms and ecosystems. Honeywell's legendary thermostat—a product designed in the industrial age, one that received several industrial design awards—is now colliding against a digital ecosystem of hardware, software, connectivity, and services with many different but plausible business models. In the industrial age, Honeywell could have developed an effective strategy for its product on its own. Now, it has to develop its strategy as part of digital platforms and ecosystems.

This is collision at the core, with Honeywell (and other incumbent competitors in its traditional industry) competing against Alphabet and its digital prowess. Honeywell could legitimately have ignored Nest in 2013, but by the beginning of 2016, the industrial world of home energy and comfort has collided with the digital world of smart homes shaped by the Internet of Things, mobile operating systems, data analytics, and cloud connectivity. In 2013, Nest was an experiment at the edge. Now it's colliding with and directly challenging Honeywell. Tomorrow, it could challenge a wider range of companies in different industries.

How has Honeywell responded? By designing a digital thermostat with the brand name Lyric to add to its portfolio.[7] For Lyric to succeed, however, Honeywell's transformation road map has to be more comprehensive than adding a single digital product, and its organizational model may need revision.

The immediate tactical question for Honeywell management was: What software operating system (OS) should Lyric run on? Initially, it chose to work on Android OS—the very operating system under the control of Alphabet that owns Nest. Then it expanded the software scope: it made its thermostat work with Apple iPhone's HomeKit, Samsung's SmartThings, and other relevant standards that are jockeying to be part of the digital smart home hubs. The selection of OS software is at the core of digital transformation not just for Honeywell but for every company whose product gets digitized and connected. So collision at the core is not a simple and straightforward case of two industrial giants taking advantage of digital trends to upgrade and redesign their respective products; it's an example of one industrial company competing against (and potentially cooperating with) a digital giant. It is at the core of the Digital Matrix, as it changes our traditional view of competition and cooperation, which I will discuss in Chapter 7.

The larger strategic question for Honeywell management is: What's your plan for connecting Lyric as part of larger ecosystems? Unlike a traditional thermostat, which is a standalone product, the Nest thermostat is part of a platform with other products from Nest and external partners that make up the Works with Nest ecosystem. And the advantage of this platform is that it can be scaled exponentially by adding new products from new partners. Furthermore, because it is digitally connected and therefore capturing data, all of the partners in the ecosystem can earn additional revenue through services based on information about their products in use. Tony Fadell may have left Nest in June 2016,[8] but Google still considers the company an important part of its road map to influence and control homes. How will the Lyric ecosystem compete and/or interconnect with the Nest ecosystem? Or will Lyric become part of the digital ecosystem orchestrated by Nest or another company?

Two important lessons arise from this example. First, every product designed in the industrial age will sooner or later compete against products designed and delivered by companies that are born in the digital age. And when that happens, the competitive landscape shifts from product-versus-product from individual companies as in the industrial age to products within and across ecosystems comprised of many businesses. Such ecosystems take advantage of developments in digital technology to redefine the very functionality of products and change the value propositions by adding services based on information and analytics around their use. In addition to merely keeping a home at the right temperature or sounding the alarm if smoke or carbon dioxide reaches dangerous levels, Nest delivers home comfort and peace of mind with real-time monitoring and cloud connectivity. In other words, *products get embedded in digital ecosystems with broader value propositions.*

Second, when value propositions change, revenue models and profit drivers change, which implies that organizational models must change as well. Nest is a company whose core competencies are hardware design, software applications, algorithms, and analytics, which are very different from those of industrial companies. Nest fits in with Alphabet's portfolio and ways of working, whereas Lyric and the broader digitization of products are alien to Honeywell and most industrial-era companies. The technology risks that Honeywell faces with digitizing its business are different from Nest's, as Alphabet is better positioned to manage such technology risks. Honeywell may have a better understanding of the distribution and marketing channel issues, but are they difficult for Alphabet's Nest to comprehend and differentiate? *Digitization redefines core organizational capabilities and ways of working.*

That's why the collision at the core of industrial companies like Honeywell and digital companies like Nest is more than an issue of technical architecture or minor changes to the product road map. The collision is about strategy models and organizational models. Incumbents must not only recognize but also, more importantly, respond to the collisions. Astute leaders must lead their organizations to embrace digitization and make the necessary changes—however tough and daunting they might be. The alternative is not so attractive.

HOW CAN YOU EMBRACE DIGITAL TECHNOLOGY AND ADAPT?

It's not that incumbents are unable to adapt. It's that ultimately your successful transformation depends on the level of commitment that your business makes to digitization. It means investing in new competencies in hardware design, software, and applications, and making new acquisitions and new sets of alliances and partnerships. Equally important is recognizing the new organizational capabilities you need to compete against companies born in the digital era.

Of course, Honeywell against Nest (within Alphabet) is by no means unique. Similar collisions are occurring in many industries. For example, senior managers in global hotel chains are facing competitive threats from Airbnb. The Accor hotels group—which has about 3,900 hotels worldwide—has responded by launching a comprehensive digital transformation strategy to position itself as a leader in "digital hospitality."[9] It has committed $250 million to upgrade its digital infrastructure to offer guests personalized service through an integrated mobile application that creates a "seamless journey" from one-click booking to electronic payment to online check-in to real-time feedback. Through its ecosystem, it is connecting customers, employees, and business partners.

So what is the appropriate response when a new business model is threatening your own? Even if you've been keeping track of experiments at the edge and carrying out tactical experiments of your own, how do you take advantage of the changing landscape rather than being swallowed by it? The experience of the auto sector can help provide some insights.

A car, or a computer on wheels connected to the cloud? Learning from the automotive sector

In the changing world of transportation, it's now fair to ask: What's an automobile anyway? Is it the industrial combustion engine that we first called a horseless carriage in the late nineteenth century? Or is it a network of computers with tens of millions of software codes connected to the cloud? Which one is a more appropriate way to think about this vehicle in the twenty-first century?

Framing a car as a physical product with services wrapped around it limits our thinking to the way the automobile industry functioned as linear value chains until the end of the twentieth century. By visualizing it as an intersection of software, entertainment, telecommunications, and cloud connectivity, we see the beginning of a mobility and transportation ecosystem where the auto sector interconnects with other industries to deliver broader value propositions around mobility and transportation. This thinking is consistent with views expressed by Bill Ford and Mark Fields at Ford Motors to look beyond the physical product to weather the collision at the core of the automotive sector. And they are not alone in this thinking.

The Consumer Electronics Show in Las Vegas is not typically where CEOs of car companies have delivered keynote speeches, but it is where Dieter Zetsche gave his first speech as chairman of the board of directors for Daimler AG in 2012. He noted, "Here at the Consumer Electronics Show in Las Vegas there are some people who view the automobile as an accessory to consumer electronics. Conversely, at the auto show in Detroit there are many people who view consumer electronics as mere trimmings for the car. Both points of view miss the point: as much as a smartphone can be far more than just a tool for communication, a smart car can be more than just a means of transportation. Precisely at the interfaces between communication and mobility, vast potential for innovation lies dormant, and we intend to tap it."[10] Daimler, the parent company of Mercedes-Benz, saw how different industries could morph into mobility and transportation ecosystems, and it had already partnered with Google to bring Google Maps and Directions, Street Views, and related search programs into the dashboards of Mercedes cars. These experiments were intended to provide complementary services to Mercedes drivers with telematics. Google and Daimler collaborated to co-create new value for customers.

During the first fifteen years of the twenty-first century (and even before), many tech entrepreneurs, digital giants, and even industry incumbents were experimenting with such digital technologies at the edge of the automotive sector. As Tony Douglas, head of strategy at BMW's mobility services unit, remarked, "Instead of just producing transport hardware... we have to get into the service industry in a larger way."[11] These services, often produced in partnership with digital giants, centered primarily on safety (monitoring airbag deployments and unauthorized door unlocks through cellular links), navigation (providing directions through voice prompts), convenience (offering access to roadside assistance through a mobile app), communications (creating an in-car Wi-Fi hotspot), entertainment (integrating music and games with streaming technology into the car's dashboard), etc., and were delivered through digital channels as a way for car companies to earn additional revenue and a higher percentage of the profits. Proprietary digital systems such as GM's OnStar, Ford SYNC, Mercedes' mbrace, and Toyota Link were also a way for individual firms to differentiate their products relative to other carmakers and to complement their automotive technology.

Recently, however, many of these experiments have started to challenge the incumbents' traditional business model. As automobiles become computers on wheels—especially as more automobiles are manufactured with hybrid and electric drivetrains—Google and Apple have sought to have greater influence. Google has extended the scope of its Android software beyond mobile phones to cars and beyond dashboard maps with its Android Auto initiative. This mobile app offers drivers control over GPS mapping/navigation, music, messaging, phone calls, and web search, using touchscreen, button controls, or voice commands. And it is compatible with a variety of mapping and music apps. Likewise, Apple has extended its iOS beyond iPod integration to automobiles with its CarPlay app, which offers similar controls and is also compatible with a wide range of other apps. These two digital giants have courted all the major carmakers to be part of their expanding ecosystems with software at the core.

More importantly, Google has successfully demonstrated its self-driving car project. Its cars have self-driven more than 1.5 million miles,[12] and the company has received a key patent[13] relating to autonomous drive late in 2011. Google and Fiat have entered into a preliminary agreement to collaborate on next-generation automobiles.[14] And Apple is rumored to be working on its own secret car project while investing $1 billion with a Chinese car-sharing company, Didi, a competitor to Uber.[15]

By early 2016, the collision between traditional automakers and digital giants was plainly obvious to everyone. Zetsche commented on the "opportunity from the convergence of the West Coast technology and the auto industry with its huge technology depth... We are not afraid. We are confident about our own strength."[16] At the same time, a Toyota senior executive remarked: "Carmakers don't want just to become a kind of commodity, where we will only deliver an empty box and somebody else will put in the box something which will be the real added value."[17] And Tony Douglas of BMW remarked, "The transportation industry is ripe for disruption. Either we kind of drive that disruption and gain from the new business models that will emerge, or we let someone else do it."[18] These three comments clearly capture the tension in this phase of digital transformation. If digital competencies become the differentiators and carmakers do not excel in them, they could be pushed to the commodity end of the ecosystem (just as happened to manufacturers of personal computers and mobile phones as they became part of broader software ecosystems).

We may not yet know the dominant operating systems that will drive the automobiles of 2025, just as, back at the turn of the twenty-first century, we could not have predicted the operating systems that run our mobile phones today (Apple and Alphabet had no presence in mobile telephony at that time). But we do know that digital giants are capable of strongly influencing this industry as it races towards digitally defined multimodal transportation. At this juncture, then, the executives of incumbent auto companies face two decisions.

1. Should automakers attack with software?
   Traditional automakers are acutely aware of their own limitations with software competencies when positioned against Google, Apple, and Microsoft. But the strategic question facing every automaker is: Could cars go the way of personal computers and mobile phones? That is, will the value shift from hardware to software in cars, as it did with personal computers (we went from calling personal computers IBM PCs in the mid 1980s to Windows PCs by the mid 1990s) and mobile phones (we went from feature phones controlled by Nokia and BlackBerry to smartphones running on Google's Android and Apple's iOS)? With cars

becoming computers-on-wheels-connected-to-the-cloud, how should carmakers defend their distinct role and unique value? They all recognize that they have to transform their core business model to become digital, but the question is how? Do they invest in their own software systems? Partner with tech entrepreneurs or digital giants to co-create new forms? I will explore such inevitable tensions as I develop the winning moves in Chapters 6 and 7.

2. Should automakers defend with apps and app stores?
   Even if the auto operating systems are under the purview of someone else, could auto brands control the app stores and thereby generate revenue and ensure profits by providing service and ensuring customer loyalty? Should automakers create car-specific app stores or let the apps be defined and curated by the digital giants? At this point, Ford has defined a three-level architecture that links a) apps that are built into the car in the factory, b) apps that are installed by the dealers to provide location-defined services, and c) apps that are introduced by the car's drivers and passengers.[19] Increasingly, the collision will focus on who gets to control the app architecture and stores as well.

The automotive industry illustrates the kinds of questions that will apply to nearly every industry as products get digitized and apps become central to connect and interoperate products on networks. The key collisions involve the processes and practices of how industrial-age companies function. As an industry incumbent, your challenge is to rethink your traditional product-centric decisions against the broader landscape of industry solutions that may become more central in the coming decade. Your challenge is to learn to work cooperatively and competitively with other incumbents, and with the tech entrepreneurs and digital giants.

Even if your industry has not yet transitioned to this phase, you can and should begin to understand how and when collisions happen in other sectors and use them to anticipate what might happen in your setting. It can certainly be instructive to look at industries similar to your own, but it can be just as valuable—and sometimes more so— to look further afield. Ultimately, however, you will have two levels of response.

TWO LEVELS OF RESPONSE: COEXIST AND THEN MORPH

When digital business models first appear on the scene, they are fuzzy, ill formed, and inelegant; most incumbents do not see their differentiated value or perceive their competitive threats. Indeed, research over two decades has shown that incumbents do not recognize disruptions and, consequently, fail to develop effective responses. How did Walmart respond to Amazon when it first introduced e-commerce? How did Barnes & Noble respond to Amazon? How did Blockbuster respond to Netflix? How did Britannica respond to Microsoft's Encarta, and later to Wikipedia? How did BlackBerry and Nokia respond to Apple's iPhone? How did Microsoft react to Apple's iPhone launch? In all these cases, the incumbents failed to recognize the gravitas of the digital innovations; they did not decipher the intensity of would-be collisions. Now that you have been schooled on theories of disruption[20] from the periphery, it is my hope that you will not miss such potential threats from tech entrepreneurs and digital giants. In my view, however, even if you recognize the threats, you may still have problems in figuring out effective timely responses. Here is some advice.

COEXIST

The best initial response when you recognize potential challenges from digital upstarts is to set up a unit charged to create an effective digital model to compete. So, in essence, you are setting up digital and traditional business models to coexist within the same structure. In the case of Netflix, the old DVD-by-mail model and the new video-streaming model coexist as Reed Hastings and his team figure out the best way to phase out their historical business model. There will, of course, be the usual challenges: insufficient resources allocated to the digital units to develop and grow, inadequate digital capabilities to effectively compete against newer startups and digital powerhouses, diffused management attention across traditional and new units resulting in neither succeeding. These challenges need to be managed.

While you are incubating a digital business within the traditional structure, examine the following three avenues:

Sharpen your value-to-cost advantage. Digital business models do not start off being superior along all dimensions: there's the unfamiliarity, steep learning curve, poor links with other products and services, etc., to overcome. As a result, some customers approach innovations cautiously as they wait for the initial bugs to be ironed out. So sharpen your traditional cost-to-value advantage using the types of analysis underlying, for instance, the "blue ocean strategy."[21] Instead of trying to beat the competition, create value in untapped markets by simultaneously pursuing differentiation and low cost. Instead of having your industry shape you, make customer value your primary focus and shape your business accordingly. Examine how venture capitalist Geoffrey Moore's ideas of the hierarchy of powers[22] reinforce and delineate your offering in the marketplace. In other words, think like an investor. Start by assessing the power of your industry, then hone in progressively tighter on your own power within that industry, within your particular segment of the market, among competitors producing the same product or service, and in situations that favor no company in particular. For example, Honeywell might look at ways to use its traditional thermostats to differentiate itself from competitors and create lower-cost value for customers in currently uncontested market space. Traditional automakers might use this same logic. And traditional hotel chains might benefit from more sharply framing their distinctive features and superior services. Some formats of Accor hotels, for instance, offer no restaurants or lounges; no desks, closets, or drawers in the rooms; no staff in the lobby, but their beds are as comfortable as those in two-star hotels for half the price.

Examine your alliance advantage. Think about the ways you can use alliances to defend your traditional business while you examine new options for the digital business. For instance, three German automakers—Daimler, Audi, and BMW—pooled their resources in August 2015 to buy Nokia's HERE maps, possibly as a defensive measure against other automakers acquiring this technology.[23] When the Apple Watch served notice to the traditional watch industry that digitally connected watches had become a reality, luxury watch manufacturer TAG Heuer joined with Google (Android) and Intel to co-create a connected watch as a complement to its traditional line.[24]

Evaluate your acquisition advantage. Often tech entrepreneurs, especially the ones that challenge your traditional model, are interested in being acquired because they may not have the skill or the resources to scale up beyond the initial threshold funded by venture capitalists. Some prefer this route to going public with an IPO, and your scanning activity in phase 1 will prove valuable in identifying potential acquisitions. You may still want to keep these entrepreneurial units separate from your core business to allow them to function with autonomy. Walmart's e-commerce initiative, which goes under the @WalmartLabs umbrella, knits together dozens of acquisitions, such as Kosmix, Vudu, Grabble, and Yumprint.[25] It coexists with Walmart's core big-store operations and associated acquisitions.

Consider this coexistence stage a wake-up call. You want to clearly and sharply delineate the distinctive value propositions of your existing models to preserve your revenue and profits. At the same time, recognize that the digital alternatives should not be underrated or minimized. Maintain your traditional business model at the core, but experiment with innovative digital alternatives as important satellites. In this stage, then, traditional industrial business models coexist alongside new digital business models, perhaps within your company but certainly within your industry. These physical and digital spheres complement each other, and some companies allow the two to coexist for a long time; there's really no predefined time frame in which traditional models fade away and newer digital alternatives take hold. It's up to you to decide strategically when to scale up the digital part of your business and cross over from physical to digital—or, as I call it, when to "morph your core."

MORPH

Coexistence of dual business models is temporary. Like every product, service, and technology, every business model has a life cycle. When depicted visually, this life cycle looks like an S: one end marks its birth (or creation), the middle section shows its growth and adoption, and the other end marks its death (or obsolescence). Usually, the end of one S-curve indicates the beginning of a new one. In other words, in the early stages of digital transformation, the traditional core gets more attention than the digital counterpart, because it is more familiar. But over time, as you decide to commit more resources to your new digital model, the core of your business will morph from traditional to digital. Historically, incumbents have missed this transition point, because the bends in the S-curve often appear fuzzy when looking ahead. Only in retrospect can we clearly see the missed timings.

So to make the best use of the window of transition while you are morphing your core, follow these three principles:

Divest from traditional businesses to focus on your new digital core. This phase is about moving the center of gravity of your business away from the industrial age. An effective strategy is to rebalance the scope of your business instead of simply adding a digital component. Think about what assets you can divest to focus on managing future possibilities rather than current problems. For instance, GE shed its finance arm and other non-strategic assets to focus on businesses aimed at increasing efficiency in transportation, buildings, power, and health care through digital technology and data analysis. Similarly, IBM shed its low-value businesses, including personal computers and point-of-sale terminals, to focus on newer areas such as cognitive computing and the Internet of Things. Look at cloud service providers such as Amazon, Microsoft, and IBM as possible options for the digital core where your internal capabilities may fall short in the beginning. If divestment is not a timely option, consider separating your traditional assets so that you can focus on the ones most likely to form the basis of your reinvention in the next phase.

Absorb digital acquisitions into your core. During the coexistence stage, many companies undertake acquisitions to get started with digital capabilities and they often keep these businesses separate. In this second stage, it's important to integrate these firms into your core business. Doing so will force you to learn from their digital business model and drive your own internal strategic and organizational change. In other words, to reap the benefits of acquiring Cruise Automation, a tech entrepreneur with competence in autonomous vehicle technology,[26] GM must now absorb the startup's functionality and challenge its newly created engineering team dedicated to autonomous driving to redefine and evolve the core of the company itself. Similarly, Daimler, Audi, and BMW must now integrate the HERE maps' functionality into their own automobiles in ways that differentiate them from the digital giants' automotive mapping technology. And Toyota must assimilate the sixteen engineers it acquired from Jaybridge Robotics in 2016 to accelerate its digital transformation.[27] Similarly, Monsanto recently acquired the Climate Corporation, a digital agriculture company that analyzes weather, soil, and other data to help farmers maximize their yields.[28] So far, Monsanto has tried to keep Climate's operations separate from its own, but it must soon integrate the two and put data science at the core of the entire company as it positions itself for reinvention.

I expect to see a more sustained pattern of acquisitions and alliances between traditional industry incumbents and tech entrepreneurs so that the incumbents are able to morph their core and stand toe to toe with the digital giants in the next phase. To be prepared, you must put digitization at the center—which brings us to the next principle.

Instill digital business thinking at your core. To truly morph your business, you must shed your traditional way of thinking. Instead of thinking about your product, your service, your industry, your individual business units, you need to think more broadly. You want to be thinking about who you can work with to create products and services that bring value to consumers in ways and in places that you—and they—have never thought of before.

Consider the *New York Times*. Many of its traditional newspaper competitors have already ceased to exist, and the *Huffington Post*, a digital news source, has overtaken the *Times* in reader traffic. So how's this legendary newspaper responding to the collision at the core of its business model? An introspective analysis concluded that although the newspaper has been winning at journalism, it has lagged in "the art and science of getting our journalism to readers ... We haven't done enough to crack that code in the digital era." So management then charged a group of employees to come up with specific recommendations about the company's digital innovation. Their report asked management to "reassess everything from our roster of talent to our organizational structure to what we do and how we do it," to "rethink print-centric traditions, use experiments and data to inform decisions, hire and empower the right digital talent." In other words, it has to reimagine: What is page 1 on the Web? What is it on the mobile web, where individuals consume news differently? What does it mean on the social web, where the importance of news is socially curated? The *New York Times* is challenging its historical print-centric traditions and reimagining the newsroom of its future.[29]

And Monsanto is morphing as a leader in data science–driven agriculture (often referred to as precision agriculture) by providing farmers with recommendations based on data analytics of the kind more commonly found on Wall Street and in Silicon Valley. The idea is to combine data and analytics to minimize risks and maximize yields by taking into account weather challenges, soil health, weed management, insect profiles, and disease characteristics. That is the scope of digital thinking needed when strategy *and* organizational models collide. Anything less will be inadequate and ineffective.

Over time, the hybrid model of traditional industrial and newer digital models coexisting will prove limiting, with competing pulls and misalignment with market demands. In due course, successful organizations will effectively morph into a fully digital business model—in every industry over time. Once you have shed inefficient ways of working, you start to resemble the tech entrepreneurs but with scale. You start to see the power and capabilities of digital giants. And you begin to see a future in which your domain knowledge and detailed subject in your industry combined with digital thinking will give you as much right and competence as the other players to be the leader in the next phase. You well know that you must go beyond just overlaying digital technologies and morphing your core. The key to stepping up is absorbing today's digital functionality in order to reinvent your business at the root, which is the subject of the next chapter.

CHAPTER 5

REINVENTION AT THE ROOT

{re • in • vent | to replace something with a new version}

"**W**HAT BUSINESS ARE you in?" No doubt you, like several generations of managers, have been asked that question about your own business or about others. Ever since the *Harvard Business Review* published Professor Ted Levitt's famous article "Marketing Myopia" in 1960,[1] that question, or a variation of it, has been posed in classrooms and meeting rooms worldwide. But have you really thought about that question in relation to digital technology, against the backdrop of pervasive digitization in your industry and your business itself?

"WHAT BUSINESS ARE YOU IN?"—REVISITED

By now you've realized very vividly that when your traditional business model collided with encroaching digital business models, that legacy model could only survive for so long. You know that to stay relevant, you need to reinvent the root of your own business. Looking at the business logic that others are inventing can help you with your own: What are tech entrepreneurs innovating in other industries? How are other incumbents reinventing their business models? Your focus is on trying to understand the compelling logic of digitization at the core of your offering. On your Digital Matrix control panel, the three screens in the right-hand column represent the various reinventions being carried out by the three sets of players. As your industry changes, they are all asking themselves: "What business am I in?"

The classic debates were: "Is GM in the industrial combustion automobile business or in the transportation business?" or "Is the *New York Times* in the newspaper business or in the news business or in the publishing business?" or "Is IBM in the IT hardware business or in the software and services business?" or "Are Coca-Cola and Pepsi in the nonalcoholic beverage business or something different?" And the answer to all these questions in the digital era is that new business models emerge at the intersection of industries. An organization must learn to think of itself not as producing goods or services but as doing the things that will make people want to do business with it. So GM's business is not narrowly defined as designing and manufacturing cars with internal combustion engines. And it's also not just a company that finances or services these automobiles. In the digital age, GM must think about its role in the changing landscape of multi-modes of transportation that include both products and services. Cars with internal combustion engines are its legacy, but they may not be its future. The *New York Times*, IBM, Coke and Pepsi, and nearly every company will need to redefine and reinvent their businesses.

To do this, you need to focus *not just on the design and delivery of products and services but on solving problems and shaping solutions*. We see the problem-solving focus in digital giants and entrepreneurs—so let's look at one illustrative case each below.

What's the big problem? Learning from Facebook

So what's the big problem that Facebook's trying to focus? Its mission is "to give people the power to share and make the world more open and connected."[2] In giving power to the people through tools (Facebook profiles, pages, apps, Messenger, and so on), the digital giant has created ways for brands to have direct and individualized links to customers, and the company is able to leverage this conduit for its revenue and profits. There is no predefined industry boundary that limits how Facebook thinks about the power of sharing to people across the globe and helping the world become smaller than the familiar "six degrees of separation." Its WhatsApp and Messenger apps aim to redefine interactions between consumers and businesses (conversational commerce) using social web and messaging platforms. Its Facebook Live video-streaming app connects broadcasters and viewers in real time and competes with YouTube, Amazon, Comcast, and Verizon. Its Oculus virtual reality headset—a new tool for immersive, shared social experience—allows viewers to go inside their favorite games and movies and competes with Google (Daydream) and Microsoft (HoloLens). It is working on deploying solar-powered drones to deliver Internet access to remote rural parts of the world. In this domain, it is pitted against Alphabet and Amazon. Remember, these are no longer products; they are tools that enable individuals to share and connect on the social web. In solving the thorny frictions that prevent seamless connectivity, Facebook is not only better linking companies to customers but increasing its revenue and profits.

What's the big problem? Learning from Tesla

And what's the big problem that Tesla is trying to solve? You might expect the tech entrepreneur's focus to be about next-generation cars or electric cars with greater reliance on software and cloud connectivity. But you'd be wrong. If you asked Elon Musk to answer that question, he'd say simply: "We accelerate the world's transition to sustainable energy."[3] So is Tesla in the smart grid business? Is Tesla in the lithium-ion battery business? The answer is yes to both questions. But it is not about delivering products and services in the ways past. It's not about defining the business to answer Professor Levitt's question in the 1960s. It's about combining batteries with fast charging stations to play an integral part in shifting from gasoline-powered modes of transportation to electric (and more energy sustainable) ones. Tesla's lofty goal lies in solving problems at the intersection of industries—energy, transportation and mobility, and home comfort. In trying to solve this big thorny problem of sustainable energy, Tesla has been working with Panasonic to produce more lithium-ion cells by 2020 than all of the world's combined output in 2013.[4] The joint venture operates a gigafactory that also produces battery packs for use in stationary storage in homes and businesses, which improves the robustness of the electrical grid, reduces energy costs for businesses and residences, and provides a backup supply of power.

So is Tesla an automaker like GM or Ford or BMW? Yes, in the conventional sense, because it designed and develops cars, just like these companies do. No, because Tesla provides electrical charging at home and in a network of super-charging stations, but the traditional automakers do not provide free gasoline! In Tesla's own words: "Tesla is not just an automaker, but also a technology and design company with a focus on energy innovation."[5] But energy innovation is not a business with defined competitors and standard definitions of industry boundaries. Today, Tesla makes electric cars and batteries, but its business scope could expand and morph with different pieces focused on accelerating the transition towards sustainable energy. Viewing Tesla in terms of products and services delivered (and the markets served by them) misses the company's larger purpose to influence the rules at the intersection of the above industries.

So do Facebook and Tesla have anything in common? They define their mission in terms of solving problems, unconstrained by conventional industry boundaries, product/service definitions, or revenue models. They take advantage of powerful digital technologies to address their ambitious missions. They bring their digital prowess into traditional industries. They seek to define the rules of value creation and capture when big problems are solved more effectively than ever before by bringing the power of digital technologies to bear. And over time, they are likely to evolve their domains of problems to be solved.

WHAT THORNY PROBLEM COULD
YOUR BUSINESS DISTINCTIVELY SOLVE?

Instead of "What business are you in?" the two fundamental and inter-related questions you need to ask as you reinvent are:

1. "What *problems* are we trying to solve for whom in the world?" and
2. "How are we *uniquely solving* them by taking advantage of digital technologies?"

These two questions compel you to look at the ways digital technologies have altered the core elements of the business model and shifted the activities from one company to another in the broader ecosystem. They also compel you to examine how new interactions could be effectively monetized. The key point is that such problems have to be solved at scale, across industries, and at speed. These are not high-touch, high-fee, personalized, bespoke solutions for a few wealthy segments; these are high-value, highly innovative solutions for large segments of society. So instead of defining the business you want to be in, focus on the thorny problems that you want to solve. Examples from three industries can help guide your thinking.

Health care. Novartis is a Swiss pharmaceutical company whose chief executive officer, Joseph Jimenez, has been on a mission to focus "beyond the pill" and solve core health care problems. In his words, "I really believe that in the future, companies like Novartis are going to be paid on patient outcomes as opposed to selling the pill."[6] He has realized that delivering pharmaceuticals without guaranteeing their effectiveness as part of a patient's overall treatment may have worked in the past, but present and future consumers are focused on information, analytics, and holistic approaches to health and wellness. They want to know the complex interactions between treatments and outcomes, and they want health care personalized for their individual needs. Pharmaceutical companies like Novartis that reinvent their business model to solve problems are seeking to differentiate from those in the segments defined around simply producing pills.

Industrial Internet. GE, too, is seeing bold new possibilities for problem solving. Long known as a home appliance company, it is reinventing itself by combining its traditional competencies in materials science and physics with new capabilities in data, software, and analytics to frame and effectively solve core problems in four sectors, as we saw in Chapter 4. For example, by using machines with sensors and software connected to the network, GE can observe, use predictive analysis, and provide proactive solutions to enhance efficiency in factories. With real-time visibility, the end-to-end manufacturing processes can be tightly controlled, and maintenance can be accurately predicted and carried out when it's needed and when it's convenient rather than on a predetermined schedule. So GE solves its customers' "inefficiency problems" in ways that were impossible without digital technologies.

Other companies in the industry such as Bosch, Siemens, and Samsung are now also using data and analytics to find and eliminate their own pockets of inefficiency. The result, which the German manufacturing companies call Industry 4.0,[7] is an increase in productivity and a decrease in cost and time delays across different companies in the value chain. Studies by GE, McKinsey,[8] the World Economic Forum, and others believe that such reinvention at the root of manufacturing and supply chains could save many billions of dollars over the next decade. Suddenly, the invisible domains of digitization—manufacturing, supply chains, and distribution—become as important as the visible domains of digitization—the customer-facing interactions with apps, ads, and algorithms.

Business solutions. IBM's reinvention under chief executive officer Ginni Rometty marks a significant shift from its historical role as a *systems integrator*, meaning that it tied together the disparate pieces of systems (computing hardware, software, and services) better than other competing integrators or than its customers could do by themselves. The reinvention, in my view, positions IBM at the intersection of technology and deep domain knowledge to address questions such as: How could cities function more effectively given massive urbanization? How could quality health care be delivered affordably in critical areas such as cancer? How could blockchain technology be used to create more trust? How could big data and analytics seamlessly enhance knowledge work? IBM is positioning itself as a *solution integrator*, meaning that it brings relevant knowledge based on its extensive experience across industries (its wide scope) to uniquely solve each client's unique business problems.[9] As we saw in Chapter 4, IBM, under Rometty, has divested the company from its low-value hardware products to focus on high-value areas, accelerated its investments in cloud and artificial intelligence technology, and created a separate unit to commercialize the cognitive computing frontier under the Watson brand. It has also made significant acquisitions, such as the Weather Company[10] and Truven Health Analytics,[11] to shore up its domain knowledge in specific verticals and created global alliances with companies such as Apple,[12] Microsoft, Cisco, and VMware to expand its scope and scale of expertise.

Precision agriculture. Monsanto's reinvention is to go from supplying genetically modified seeds to solving the problem of how best to maximize farmers' yields. This essentially calls for combining traditional competencies in agriculture seeds, seed treatments, and crop protection treatments with digital competencies in accessing data from multiple sources and analyzing it to offer precise solutions (such as changing the planting calendar or irrigation schedule). In selected tests, Monsanto has already shown that by varying the amount of seeds in different conditions, it could help corn farmers improve yields by about 7 percent.[13] To further ensure that it leads along this trajectory, it has established Monsanto Growth Ventures, a venture capital group mandated to invest in entrepreneurs dealing with emerging areas such as software, remote sensing and measurements, robotics and automation, biotechnology, and novel business models in agriculture. The company is laying the foundations for reinventing its business model.

WHAT BUSINESS MODEL IS RIGHT FOR REINVENTION?

I've said that digital transformation involves a shift in thinking that leads your business in three phases from a product- or service-focused business model to a problem-solving one. To unpack this shift, it can be helpful to look at four archetypical business models (see Figure 2), starting with two that will already be familiar to you.

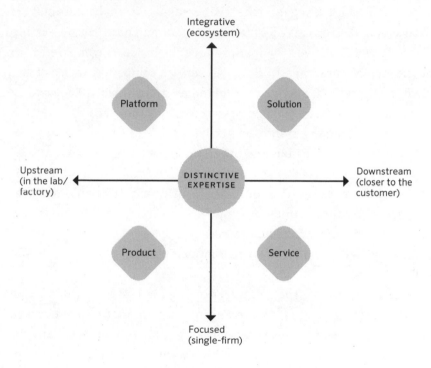

FIGURE 2 Four Digital Business Models

Product. These are the tangible items offered for sale that we know from the industrial age. They're computers, refrigerators, washing machines, tennis racquets, and lightbulbs. In the digital era, companies still make the same products, but they are fitted with sensors and software to capture data and link them to other products and to services. Products get smart as they are digitized; for example, cars, as we've seen, started with telematics and have now become computers on wheels connected to the cloud.

Service. These are the intangible products, or the actions that are performed to fill a need or satisfy a demand. We're familiar with services such as hotels, banking, entertainment, and education from the industrial era, too, but digital-era services are supported, shaped, and delivered by digital technologies. Here again, services get smarter as the data about them gets richer. Whereas traditionally automakers had a lot of data on cars up until the point of purchase (the ownership record), which allowed them to produce the best cars compared with their incumbent competitors, newer service companies such as Uber, Lyft, Didi, and Waze define their mandate more broadly. They collect reliable, detailed data on how cars are used (driving record) or related pain points (parking, traffic) or auxiliary services (insurance, maintenance, or reliance on taxis and other modes of transport) so that they can deliver the best transportation service from A to B at specific times. This more detailed "outside-in" information about customer expectations and experience is much richer than the old "inside-out" information derived from market research data.

In the digital context, two other models emerge.

Platform. These are the computer operating systems, video game consoles, smartphones, search engines, and more that have arisen in the digital world. As we saw earlier, platforms connect many different types of companies transacting with many different types of customers. Or as economists David Evans and Richard Schmalensee note in their new book *Matchmakers*, platforms "are inherently multi-sided because they provide physical or virtual space for two or more groups to get together."[14] As platforms grow by digitally connecting more individual companies, the value to consumers of their products and services is increased because they work together in a way that none could achieve on its own. In the case of the automotive industry, these are Apple CarPlay and Android Auto for now. They now seamlessly extend services from smartphones to cars, but their scope could expand further in the future.

Solution. These customizable products, combinations of products, services, or mix of products and services have become more prevalent in the digital world, in which companies are better able to solve specific business problems through data and analytics. For example, we are on the cusp of seeing digital technologies, such as the conversational bots we saw in Chapter 3, that observe and understand you and then recommend precisely what you need, delivering a personalized solution at an affordable price. In the automotive sector, GM's Maven is beginning to offer personalized car-sharing services.[15] Daimler's Moovel is reinventing the future of urban mobility, essentially using technology to connect modes of transportation so that individuals can travel in different ways, including walking, biking, riding public transit, and driving, based on what's convenient and easy to use.[16]

The reinvention at the root of the auto sector brings all four of these business models into focus. Inside-out thinking is focused on design competencies such as the styling and shape of cars, whereas outside-in thinking is focused on problem-solving competencies such as how best to go from A to B given specific individual preferences and constraints. The former excels in assembling the required suppliers and subcontractors to streamline the supply chain and make it efficient for producing the best car at the target price, whereas the latter excels in constructing a network of options to solve transportation problems for customers. The latter both expects and delivers a wider range of options that go beyond cars and could include shared and public transportation of different kinds. For example, a user may prefer the fastest transportation option at one time and the one with the smallest carbon footprint at a different time. Or she may prefer to use personal transport at one time and shared transport at a different time and, soon perhaps, a self-driving vehicle at one time and a human-driven vehicle at a different time. The key is that the underlying data, information, and knowledge in the two ways of approaching the problem are different: outside-in thinking requires new sets of expertise and also new sets of partnerships, which is why we need to look beyond the two traditional business models.

Learning from the auto sector

Each of these business models has two dimensions that define a company's distinctive *expertise*, and each of these dimensions lies along a continuum. The first is the location, either *upstream* (in the lab, in the factory, away from the customer) or *downstream* (in the field, at the point of purchase, or embedded within the customer's location). For example, car companies have primarily followed the product business model and gained their distinctive edge upstream, away from consumers. In contrast, mobile car-sharing companies such as Uber have been focused on the service and solutions models and have gained their distinctive edge downstream, interacting with consumers.

The second dimension is the focus of expertise. For instance, car companies have traditionally focused only on their own individual product or service business model to gain a distinctive edge as a provider of the best luxury cars or the most rugged sport utility vehicles (*single focus*). In contrast, Apple CarPlay gains its distinctive edge by highly integrating its business model within and across ecosystems that include a wide range of players, from hardware manufacturers to car dealers to e-commerce retailers (*integrative focus*). The reinvention of the automotive industry will be a shift away from individual companies using exclusively a single product- or service-focused business model towards ecosystems of companies playing across the four different business models to solve big thorny transportation needs.

As it reinvents the root of its business, how might GM answer the questions about the problems it is solving and its unique advantage? Obviously, there are lots of possible answers for different consumer segments, but one of them might be:

1. We are solving the transportation problems of individuals in cities (*who*) during rush hour (*what*).
2. We are offering a portfolio of options, including fractional ownership, car sharing, car pooling, rides on demand, and multimodal transport (*how*).

That's a very different answer than it would have given when it was focused on the industrial economy, and it's likely that those answers will shift in the future. But for now, they're helpful in answering my initial question: "What business are you in?" And they clearly point at GM's need to obtain different data so that it can understand the patterns of usage of its cars by customers (rather than the buying patterns of car drivers) and set new performance metrics that focus on revenue and profit per customer, or share of wallet of transportation spend (rather than the unit sales of cars and trucks). The new business model requires a micro-level understanding of how different individuals move from point A to B and an ability to knit together the best set of options at acceptable price levels. And its success is measured in customer retention and stickiness that allow GM to gather invaluable data that could be mined for detailed insights. To do this, it has to assemble a network of partnerships, and data tracking and analytic capabilities, to then deliver the best option to individuals at scale and speed.

Mary Barra, the chief executive officer of GM, is leading the reinvention at GM: "We do believe the traditional ownership model is being disrupted. To what extent, no one really knows... We're going to see more changes in the next five to ten years than we've seen in the last ten years... People are still going to need to get from point A to point B, whether they are going to do it in a traditional ownership model or with sharing. For all of that, we are expanding our relationship beyond the car."[17] She has realized that GM's reinvention has to start on her watch and knows well the magnitude and urgency of this transformation.

What the automotive industry shows is that digital giants and tech entrepreneurs do not automatically have an advantage over traditional incumbents as this industry digitizes. Certainly, Google has the ambition and the capability to influence the automotive sector with Android Auto and related technologies, including electric and self-driving cars. And it's true that Uber has amassed an enormous amount of data on the riding habits of individuals, and that its app is deeply integrated with related services (Open Table, MasterCard, hotels, etc.). It's also valid that Elon Musk's Tesla has made bold innovations when it comes to sustainable transportation and could be seen as a major influencer when it produces 500,000 cars annually at a reasonable price point of around $30,000 each, supported by lithium-iron batteries and a vast network of superchargers. Even with all these initiatives, the reinvention of urban transportation is very much in its infancy. It is not yet clear who—the digital giants, the tech entrepreneurs, the industry incumbents, or some combination of these through acquisitions and alliances—will set the new rules of engagement to solve the transportation needs in the world's megacities. As Barra said: "The whole senior leadership knows there is this incredible core business that we've worked hard to make efficient and we'll continue to do [that] and win. But there are also newer models of transportation that we've got to quickly seize because there are assets that the traditional automakers have that [Google or Apple] don't have, and there are specific assets at General Motors—like our embedded connectivity—that I am confident will drive the future."[18]

All of the traditional automakers are reinventing their business—to varying degrees—because they recognize the discontinuity and disruption facing their companies and their industry. They can already see that if they fail to transform, the share of value that goes to the automobile (physical car) as part of the larger transportation and mobility solutions will decline. The physical hardware could become a commodity and the value might migrate elsewhere in the business system. This shift is precisely the reason underlying Ford's portfolio of experiments, which include offering shared cars on demand rather than by reservation for trips in the city (city driving on demand); allowing drivers within a select group to swap cars with each other (car swap); testing a dynamic on-demand shuttle bus service that picks up and drops off six to eight commuters at user-defined locations, taking the most suitable route for all passengers on board (dynamic social shuttle); providing data-driven insurance rates based on detailed analysis of vehicle performance under different conditions; and introducing a plug-in device on the dashboard to spot open parking spaces (painless parking).[19] Ultimately, the reinvention is about deciding to select one of the four models in Figure 2 and entering into relationships with businesses following the other three models so that, together, you have a compelling offering that provides superior value superior to others.

Go ahead and map your business along the scheme represented in Figure 2. What's your model today and what's your desired model in the near future?

HOW DO YOU STAY RELEVANT DURING THE REINVENTION?

How and how quickly your company and your industry transforms will be different for everyone. Typically, phase 1, experimentation at the edge, can last between three and five years, which gives most senior managers enough time to provide guidance to think about a portfolio of experiments and allocate appropriate resources. This phase is exciting because it's about the future and about the possibility of change and doing things differently. But it does not affect the current realities and can be organized separately. Your operations continue without much influence from the activities happening at the periphery, as we saw in Chapter 3.

Phase 2, the collision at the core, takes place more quickly. Companies recognize the disruptions and tension rises internally as the company makes trade-offs and tough decisions about reallocating resources between the present and the future; between the known and the unknown; between quantifiable business cases and newer, unproven options, as we saw in Chapter 4. These tussles about digital transformation usually play out over a period of about eighteen months to two years. The older, traditional models reveal their weaknesses and managers have to decide how best to morph and shift their business fundamentals to be aligned with the digital ways of working.

When it comes to phase 3, reinventing your business and designing the new rules for your company and your industry, you need to be looking at your immediate future (eighteen to thirty-six months). This phase often gets shortchanged because your focus is on maintaining revenues and profits today, but future planning is of the utmost importance to your company and requires serious attention. This reinvention also requires unlearning some of the successful modes of the past while influencing the new rules of the game. The toughest part about this phase is that even if you have a rich history of successes, even if you have survived past disruptions, and even if you have been a poster child for adaptation in the past, you need to establish your relevance in this new era of solving pressing, thorny problems. No longer can you get away with observing digital transformation or investing incrementally or coexisting. In this phase, you and your senior managers really *must* focus on digitization as a business theme. The future success of your business depends on it—and the time to act is now.

More than in either of the two previous phases, systematic thinking followed by bold actions is the key to reinventing your business for future relevance. You do not have the luxury of benchmarking many different best-of-breed cases (because they do not yet exist), yet you must frame your relevance for the digital future: you must position yourself to solve problems at the nexus of scale, scope, and speed. That means solving individual customer problems by adapting as conditions change rather than forcing predefined standard products or services to fit every situation.

As you look at your industry, you might see that the digital giants and tech entrepreneurs have pushed into the places where your products and services reigned supreme just recently. You have been an incumbent in the automotive industry for a century or more, and suddenly the attention is on Tesla and Google. You may have led the pharmaceutical industry with a string of patented blockbuster drugs, and now the focus has shifted to patients and how they respond to a system of treatments, not just individual drugs. You may have been at the head of market research for the better part of five decades, and all of a sudden you find yourself vying for relevance with the likes of Google, Facebook, Amazon, and Twitter that have amassed an unprecedented amount of information about consumers. You may have been an award-winning retailer, but now you find your customers flocking to e-commerce sites such as Amazon, Alibaba, and eBay. Again, you establish your relevance by solving a big thorny question. But which one?

A SINGLE INTEGRATED RESPONSE: PROBLEM FRAMING AND PROBLEM SOLVING, TIED TOGETHER

What's common in this reinvention phase is a shift in business logic. Incumbents are reimagining their business models beyond products and services, tech entrepreneurs are introducing new business innovations rooted in powerful technologies, and digital giants are extending their platforms and offering relevant integrated solutions. Their actions influence your own reinvention and compel you to rethink your approach to earning revenue and profits. As you determine what differentiates your company, ask yourself: What's your relevance when business logic shifts from delivering products and services to solving problems and shaping solutions?

Even if you have not fully embraced digitization beyond mobile apps and social marketing, it's time to stress test the veracity of your business logic against significant incursions by digital giants and meaningful innovations from tech entrepreneurs. Your traditional competitors, moreover, have probably recognized and responded to digitization. Your industry may intersect and interlink with related—and seemingly unrelated—industries to offer solutions that may be more attractive and valuable to customers than the piecemeal, independent offerings of the past.

It's your time of reckoning. Now's the time to take action so that you don't find yourself marginalized and relegated to the commodity end of your industry and markets. Hone in on the big problem you are stepping up to solve and the differentiated solutions you will provide.

PROBLEM FRAMING

Physicist Albert Einstein is supposed to have said: "If I had an hour to solve a problem and my life depended on the solution, I would spend the first fifty-five minutes determining the proper question to ask, for once I know the proper question, I could solve the problem in less than five minutes." Just as tech entrepreneurs found their companies around solving a core problem, your key leadership responsibility when reinventing your company is to articulate the important business (and societal) problem that you will solve with digital technology at the core. I encourage you to think about "wicked problems,"[20] those problems that are hard to solve because they are incomplete, contradictory, or changing. In his book *Wicked Strategies*,[21] Professor John Camillus documents why conventional approaches to strategic thinking and planning fail when dealing with these highly complex and very uncertain problems, and he compels managers to think of wicked problems at the intersection of mega forces without simplifying them, as most historical planning approaches have focused on piecemeal parts of the problems.

Look at your future in terms of *where your company is positioned in the sets of relevant problems and solutions at the intersection of industries.*

When managers answered Professor Levitt's question, "What business are you in?" in the 1960s, the focus was on defining the market broadly enough not to be blindsided by solutions that met the customers' needs in new and different ways. So consumers might choose cars instead of trucks, prefer flying instead of train travel, watch televisions instead of reading books, and so on.

More recently, Stanford University's design school,[22] the consulting company IDEO,[23] and others have popularized a methodology called *design thinking* that makes framing and solving problems part of the same process. In a nutshell, it involves observing and collecting data from a user's perspective before you frame your question so that you can come up with a solution based on real information. This is the key to providing value to consumers. For example, a group of Stanford students were challenged to design a low-cost, easy-to-use incubator for the developing world to help reduce the mortality rate of the 20 million premature and low-birth-weight babies born annually. After several days observing the neonatal unit of a Kathmandu hospital, the students visited several rural areas. What they found was that most premature infants were born in these rural areas and would never make it to a hospital. So no matter how good their new incubator design was, how inexpensively it could be manufactured, or how easy it was for a nurse in a hospital to use, it would not solve the *right* problem. Their real problem was to design a solution that could function in a rural environment without electricity and be transportable, intuitive, sanitizable, culturally appropriate, and, of course, inexpensive. Instead of an incubator, they developed a temperature-regulating pouch named Embrace[24] that did just that.

But design thinking should be expanded to go beyond reframing product design or service delivery. As Harvard Business School Professor Ranjay Gulati argues, digitization gives you the opportunity to deepen your outside-in thinking and put customers at the center of your business.[25] My friend and colleague Rick Chavez, a partner at the consultancy company Oliver Wyman and previously at Microsoft, is fond of getting executives to understand that the biggest challenge facing incumbent companies is one of reorienting from the familiar "product push" to the unfamiliar "customer pull." In other words, embed digital technology in your interactions with customers to better understand their needs. If you can quantify and analyze how your customers arrive at their decisions, you can better frame the right question, tailor the relevance of your solution, and succeed over those who are still telling customers what they want.

Frame the problems from the outside in. It's easy and comforting to simply define the problem in ways that match what you have to offer, but think like a consumer instead. Understand which areas your current set of products and services fail to meet. Look at where products and services from different companies create frictions and frustrations. Step into the proverbial shoes of the customer and understand their pain points, their needs and frustrations in integrating or coordinating their different daily tasks. Use a broad context, whether it be looking at a product's entire spectrum of use or an entire day in the life (DILO) of a customer to understand the depth and breadth of the problem.

Be sure when you frame your business problems that you are looking deeper than the domain of products and services to the different pain points for customers—whether they be individual consumers or enterprises. And those pain points may not fit neatly into traditional industry definitions. What if you could guarantee overall home comfort and security instead of only delivering heat and electricity or only monitoring its efficient use or only securing the perimeter? Would that provide greater value to consumers? What if you could solve the transportation needs of individuals (or families) rather than only delivering cars or arranging taxis or providing car insurance and maintenance services? What if you could deliver health and wellness in an integrated way instead of only in the ways the health care industry is organized?

Select the problems that match your passion. There are many thorny problems to be solved, but your challenge is to decide on the ones that you want to singularly focus on for the next phase of your company's evolution. Select the problem that is going to galvanize your employees, energize their passion, and unleash their expertise to solve it. In this stage, you are not experimenting, but you are committing to a course of action in which the magnitude of the problem you seek to address matches the capabilities that you have and the capabilities that you could acquire and assemble with others. Profits ultimately lie where you bring passion to solving the problems that customers face in ways that others are unable to match.

PROBLEM SOLVING

Having framed your problem, you need to determine a rational, realistic, and systematic way to go about solving it by bringing the power of digital technology to bear in ways that couldn't be done before.

Transcend industry and disciplinary boundaries. It's well known that innovations are complex, but it's also known that creative solutions emerge when you can transcend the traditional disciplines[26] and industry boundaries. Since 2000, Procter & Gamble has relied on open innovation in its Connect + Develop program[27] to tap into intellectual capital and expertise that exists beyond its corporate boundaries. Other companies have used crowdsourcing to tap into the wisdom of groups to solve problems. For example, InnoCentive invites solutions from people with different points of view in the networked world.[28] Over the last decades, digital technologies such as these have allowed solutions to be more broadly applied across industries, as the digital giants have discovered as they expand by extending their software, applications, and analytics. Or as CEO Ginni Rometty at IBM said: "I've always believed that most solutions can be found in the roots of math."[29] Artificial intelligence, machine learning, and cognitive computing—all overlapping ideas—look for patterns that lie across industries and disciplinary boundaries—that's an exciting frontier for solutions delivery with predictive analytics at the core. And this is where IBM and GE are headed in their own distinct ways. Within the next few years, more tools will be made available for many companies to incorporate new frontiers of functionality to solve business problems at scale.

Partner for problem solving. You know very well that individual companies do not solve big thorny problems on their own. You may be able to solve part of the problem by tapping into the collective expertise in your own firm, but only by working with other incumbents, entrepreneurs, and digital giants are you going to solve the larger problem. In essence, you must focus not just on *me* (your own company) but also on *we* (all of the players in your industry and other industries, too). So cultivate a network of relationships and delineate the distinct roles each player has within the ecosystem. (I'll discuss this in Chapter 6.) As you have done in the previous two phases, look at how and where you can turn your incumbent competitors (and tech entrepreneurs and digital giants) into your partners and what you can contribute to or learn from other industries. Be open to new ways of thinking about problems as you interact in the ecosystems and find creative ways to solve them.

I began this chapter by asking: "What business are you in?" How might you answer that question now? You might answer with one of the four business models shown in Figure 2, and you might note how your chosen model connects with the other three to solve broader problems in business networks. As you change your focus from a single industry to a large cross-industry problem, it's important to remember that nothing in the digital era is static. Just because you've reinvented your business doesn't mean that you won't do it again, and that watching who is experimenting with new ideas and how, and what companies are clashing with which others and how, is unimportant. In fact, you are likely to iterate and move across the three phases more rapidly in the coming years.

BY NOW, YOU'LL have begun to see the totality of the Digital Matrix. You see the individual meaning of the nine screens, but you also see how they link and interconnect. You can see that the three players and three phases intersect in complex and dynamic ways, and you can sense and understand the patterns better. You can see how they feed the scale and scope at speed that we discussed in Chapter 1. You may have sensed from many of the examples that the winners have been using some powerful strategic moves to navigate this fluid arena and succeed across the three phases. You see how the key actions flow linearly: from observe to invest in phase 1, coexist to morph during phase 2, and problem framing and problem solving in phase 3. Yet you also understand that those actions have rapid feedback (see Figure 3). You must pay close attention to the feedback across the phases, because your focus on morphing may force you to look more closely at a different set of experiments, or your reinvention as a problem-solving and solutions-focused business may begin to morph parts of your business that you had put on the back burner. These actions across the three phases are not linear; they are intertwined and give rise to strategic moves that become central and important to effectively reinvent yourself for the digital world. Three such winning moves are our focus in the next section.

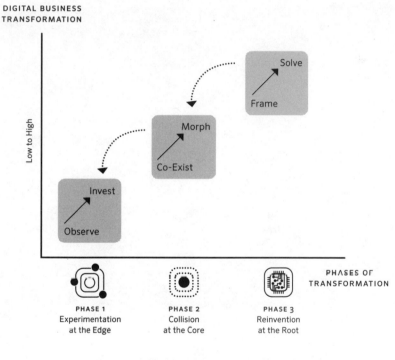

FIGURE 3 Key Actions across the Three Phases

PART 3

THREE WINNING MOVES

CHAPTER 6

ORCHESTRATE AND PARTICIPATE ACROSS ECOSYSTEMS

"It is now 1984. It appears IBM wants it all. Apple is perceived to be the only hope to offer IBM a run for its money. Dealers initially welcoming IBM with open arms now fear an IBM-dominated and controlled future. They are increasingly turning back to Apple as the only force that can ensure their future freedom. IBM wants it all and is aiming its guns on its last obstacle to industry control: Apple. Will Big Blue dominate the entire computer industry? [The audience shouts no] The entire information age? [No!] Was George Orwell right? [No!]"

SO BEGAN STEVE Jobs, Apple's chief executive officer, as he took the stage at the company's annual shareholder meeting to introduce the first Macintosh on January 24, 1984.[1] At the time, IBM reigned supreme with mainframe computers, and Apple's personal computer—developed in a scrappy garage in Silicon Valley—represented perhaps the first, if not the most important, innovation to disrupt the status quo in the computer industry. IBM's response was to create a personal computer of its own, but instead of making everything internally, as it had historically done, to speed up the new product introduction, it sought an external supplier for its operating systems (os). When IBM chose Microsoft, it did not see the company as a threat. IBM's strategy was to leverage its brand name and the power of its distribution channels to win against Apple. In reality, the competition in personal computers emerged not between IBM and Apple but between IBM PC and Windows PC. By the early 1990s, Apple computers were mostly restricted to niche segments of education and desktop publishing. In contrast, Microsoft chief executive officer Bill Gates had cleverly parlayed his initial opportunity as a software vendor on contract to IBM into dominance in the market. Not only did Microsoft supply its Windows operating system to IBM, it also sold the os to other manufacturers of personal computers such as Compaq, Dell, Gateway, Toshiba, and Sony.

MANAGE YOUR RELATIONSHIPS ACROSS ECOSYSTEMS

Microsoft's case provides useful pointers as we develop winning moves for digital business transformation for the decade ahead. It highlights two important roles in ecosystems—orchestration and participation. The first winning move involves properly managing your relationships across the different ecosystems in which you play. In industrial-age industries, the leaders are the companies with the best vertical integration (i.e., they own the key assets from end to end, whether it be a product or a service). In the 1980s, IBM was a classic example because its mainframe computer systems were made up of its own proprietary hardware, software, and services, which came only as a total package and did not work with any other company's computer system. You had to buy what IBM offered with very little modification. Or you could choose competing complete products from Digital Equipment Corporation, Data General, Fujitsu, or Wang.[2]

But in the digital age, the leaders are the companies with the best virtual integration (i.e., they control the market by assembling and managing the best network of product and service providers). Microsoft was a great example when it introduced the Windows operating system. Instead of keeping it a closed system, like IBM and Apple, they made it available to other companies' products. In other words, whereas IBM provided a complete *product*, Microsoft provided a *platform*, a proprietary wedge of software between the hardware and the different applications, to which other companies could add their products and generate their own profits. Microsoft, along with its network of hardware and software providers, developers, and retailers providing products and services for its platform, constituted an *ecosystem*.

And contrary to popular misconceptions, ecosystems do have a hierarchy, with definitive leaders and followers. I call the leaders in ecosystems *orchestrators* and the followers *participants*. Both are needed for an ecosystem to thrive, grow, and win. Orchestration is about pulling together companies with different business models and strengths in different industries and connecting them across traditional industry boundaries. This is where digital giants have intrinsic strength and superiority. Orchestration is about making links between manufacturers, service providers, platform providers, and solutions architects to create a system of network effects. Participation is about knowing your core strength and allowing others to link to you to create value greater than would otherwise be possible alone.

ORCHESTRATE TO LEAD THE ECOSYSTEM

IBM earned its revenue and profits as a systems integrator of end-to-end proprietary architecture in mainframes during the 1970s and 1980s. In contrast, by the time personal computers were widely used in homes and offices, Microsoft earned its revenue and profits as the *orchestrator* of the Windows-Intel (Wintel) personal computer ecosystem in the 1990s. Microsoft, despite controlling only a small part of the end-to-end assets in computing, shifted the locus of value from IBM to an ecosystem of complementary players that delivered a wide range of products, services, and solutions on Microsoft's platform. It co-designed the Windows platform with Intel. It partnered with Dell, which had perfected how to build to order PCs with superior supply chain capability, and even without its own physical stores had emerged as a leading reseller of the Wintel platform. It worked with Hewlett-Packard, Epson, and Kodak, whose scanners and printers provided PC users with enhanced value. In essence, through its Windows OS in the software layer and Office Suite in the applications layer of computers, Microsoft managed all of the relationships in the design, manufacture, sales, and service of PCs, from chips (Intel) to services and solutions (Accenture and EDS).

The industry's main business logic shifted from selling computers as hardware (a product) wrapped with services to selling computing solutions through platforms (with complementary products and services) for customized applications and use. Clearly, individual products (scanners, printers, software applications) and services (configuration, systems integration, consulting services) were still necessary, but Microsoft created a new pocket of value by linking them as part of broader platforms and solutions. Customers benefited, and all of the players on the platform earned revenue and profits they could not have earned on their own. Designing its personal computer operating system in ways to attract complementary players to create this system of network effects enabled Microsoft to earn both a significant share of the value as the platform's architect and the right to orchestrate the personal computer ecosystem.

Ironically, Bill Gates reached out to CEO John Sculley[3] at Apple during this period (Steve Jobs had been ousted by then) and offered to manage Apple's ecosystem just the way he had orchestrated the Windows ecosystem for IBM-compatible hardware manufacturers. That idea found no traction at Apple, which still saw its role as an end-to-end integrator; it simply was not seeing the world as platforms in ecosystems. At the time, the industrial-age business logic of integration was well understood within management circles, but the emergent role of ecosystem orchestration was relatively new, and Microsoft's golden age until the end of 2000 rested on its role as orchestrator of the PC ecosystem.

Why Microsoft failed to capitalize on its initial advantage over Windows and the personal computer to expand its ecosystems to include mobile (which Apple entered only in 2007 and has dominated over the last decade) or search (which Google entered in 1998 and has dominated ever since) or social (which Facebook entered in 2006 and has dominated ever since) is a story for another day, but it underscores the fact that ecosystems are dynamic. As we saw earlier, future success is never guaranteed by past successes. So although Microsoft orchestrated the PC ecosystem (and continues to do so even today), its role in mobile, search, social, and other emerging areas is more of participation.

In today's digital business economy, ecosystems are an essential part of doing business. Every company must now ask itself: Are we seeking to orchestrate the ecosystem or simply to participate? I will discuss how to answer this in more detail below.

PARTICIPATE TO SUPPORT THE ECOSYSTEM

Every ecosystem has several participants with different but interconnected and interdependent roles. And an ecosystem is powerful and vibrant when supported by many different types of participants whose offerings complement each other. Customers prefer these diverse ecosystems, too.

In Chapter 5, I introduced four archetypal business models. To products and services, which may be familiar, I added platforms and solutions; more importantly, I showed that each model offers a distinct expertise and delivers differentiated value in ecosystems. It is important to highlight the unique characteristics of the two newer business models. The platform model seeks to *maximize* the number of players and offerings that are linked to it by broadly interconnecting different products and services, and it seeks to increase the scale of adoption by consumers. The solution model seeks to pull together the most *relevant* set of players and offerings and integrate these specific pieces to solve the unique needs of particular customers and generate revenue and profits for doing so.

When I present this framework of business models at workshops in different company settings, I ask executives to examine in which areas their business is distinctive and in which areas they should be differentiated going forward. In many cases, these executives jump up wanting to follow all four models! I tell them that's not strategic because there are inherent tensions among the models, and they have to choose the one where they can be distinctive and invite others to work with them. Invariably, they want to become a platform architect or solution integrator, and again, I caution them against believing that these two newer digital business models are possible at all times and across all situations. *You must choose the model you are passionate about.* More importantly, remember that distinctive products and services contribute enormously to making ecosystems vibrant and valuable.

Because customer wants and needs are not constant, these business models may not all be equally important or proportionately present in every ecosystem all the time, but together, linked by digital technology, they deliver value. In other words, as a participant, you must understand that the value you deliver today may not be the same value you could deliver in the future. Choose the model wisely and evolve as conditions change.

Learning from the lightbulb

To illustrate the different ways of participating in ecosystems using these four types of business model, let's look at an example. Suppose that a light has burned out in my house. What business models are necessary to solve my problem?

Product. In most cases, a standard product (a lightbulb) may suit the purpose. Over the years, lightbulbs have been designed and manufactured in standard formats and shapes and distributed through traditional outlets. So I go to my local home hardware store, buy a lightbulb, and install it myself. The local store serves as the distribution point and the service is basic, so it does not provide much added value beyond stocking the bulbs. My problem is solved. And what do you as the lightbulb manufacturer and the retailer learn about me, your customer? Not very much, so again the value delivered is small.

Product + service. A simple—but personalized—service might be replenishing my lightbulbs through a subscription service. Here, the manufacturer or my local store may have more information about the type of lightbulb I use and how often I replace them and either alert me to the fact that it will soon be time to replace one or know exactly what I need when I phone to buy one. Still, I will buy and install the light myself, so the store provides a little personalization but not much added value. And it knows very little about me.

Product + service + platform. But let's say that I've decided to connect the lightbulbs in my home to my alarm system and my sound system via an app on my smartphone. I want my lights to turn on as soon as I disarm my alarm and open the front door, and I want my preferred music to come on at the same time. No longer is my lightbulb just a standard standalone product that operates on a standard switch. It is now connected to Wi-Fi and the Bluetooth network, and to my alarm and my music player, as part of the Internet of Things. So now I'm dealing with a product (e.g., the lightbulb from Philips), a platform (perhaps Samsung's SmartHome or Apple's HomeKit), and a service (retailing). I probably could go to my local home hardware store to buy the lightbulb, but chances are I now order it online or just press the "dash button"—a Wi-Fi device on the product—to reorder directly from Amazon or another e-retailer. I reorder the lightbulb using my smartphone, wait for it to be delivered, and install it myself. My problem is solved.

And what do you as the lightbulb manufacturer, the platform provider, or the retailer learn about me, the customer? The manufacturer learns about the usage patterns of its product, the retailer learns about my lightbulb-buying habits and can link that to information about other home products I buy, and the platform provider learns about how I use lightbulbs in conjunction with its other products and services. With all of that information, each of the providers can fine-tune its business model to give me exactly what I want and need—even if I don't know it yet. That's a significant amount of added value.

Now let's change the example slightly. Instead of replacing an existing lightbulb, I'm designing a new kitchen and need to select lightbulbs to work with the smart appliances. What business models may be necessary to solve my problem?

Product + service + platform + solution. I still need the product (lightbulb), the platform (Samsung's SmartHome or Apple's HomeKit), and the service (retailing). And I could do the research, order the lightbulbs, and install them myself. Or I could specify my requirements to a solutions architect (for example, a smart home technology designer), who will source the best individual products (the lightbulbs) with the appropriate services (for example, design and maintenance) and connections to my existing (or a new) home-networking platform. Here we've got four or more participants, with four different but complementary business models, delivering value to me, the customer. In other words, these four business models represent four ways of solving the problems that I and other customers have. Moreover, these four models are interdependent. In some cases, companies with different business models work together to create more value for the customer. In other cases, one model provides a greater share of the value. And depending on the specific situation, a company with any of the four models could play the role of the orchestrator or the participant.

I encourage you to map the roles of these business models in your setting. Which of the four models—product, service, platform, or solution—are you? How do the other three models link to you? Who competes against you for a greater share of value? How could you venture further to get a greater share of that value in the future? What model do you aspire to in the future? These answers could be different for different types of customer segments and in different settings, but you will see the roles played by the different models. Product and service are traditional business models that you can easily map, and with digitization, you begin to see the potential role of platforms and solutions: if they have not yet appeared in your business landscape, they will appear soon. The two newer models (platform and solution) complement the industrial-age models (product and service) and serve to define the role of orchestration of ecosystems.

Learning from smartphones

Remember back when the mobile world was dominated by feature phones? Nokia, Motorola, and BlackBerry were the product leaders. The telecom operators such as AT&T, Vodafone, Verizon, Telefonica, and NTT DoCoMo delivered the services. Symbian was a prominent platform, and it coordinated the development of the software on behalf of members of the ecosystem, such as Nokia, Sony Ericsson, Visa, AT&T, Vodafone, and China Mobile, as well as semiconductor manufacturers such as ARM and Broadcom. At that time, in most markets around the globe, the telecom operators orchestrated the ecosystem: they dealt with the handset manufacturers and decided which ones to promote, they managed the customer relationships, and they were responsible for the installation and upgrading of the cellular networks and for service quality and coverage.

When the mobile landscape shifted to smartphones in 2007, Apple seized control of the ecosystem. It put its proprietary iOS software at the center of its mobile broadband ecosystem and took over the customer relationship through its app stores. Telecom operators would still provide the cellular service (mainly because they owned the licensed spectrum in most countries) and app developers could continue to produce apps, but they could only reach Apple's iPhone customers on the iOS platform. Google then introduced Android in 2008 as an alternative to Apple's iOS and successfully created its own ecosystem. Not only do Apple and Android dominate their respective ecosystems with nearly 3 billion devices between them (Android clearly ahead of Apple in terms of the number of devices), they have earned the right to orchestrate those smartphone ecosystems for now. The telecom operators have ceded that role.

Samsung's role in smartphone systems helps us to understand the dynamic tension between orchestration and participation. Samsung, which competes against Apple with its high-end Android products and could be considered a co-orchestrator of the Android ecosystem, has been designing its own mobile operating system called Tizen to try to create a legitimate third ecosystem. So far, app developers haven't written prominent and exclusive apps on Tizen (as opposed to Google's Android). In the absence of network effects supported by participants, Samsung has not been able to shift its business model from product to platform. Consequently, it continues to participate in the Android ecosystem instead of orchestrating its own Tizen ecosystem.

What smartphone ecosystems illustrate about orchestration is how dynamic it is and how different the rules are from traditional business models. Past success in orchestrating another business model (service provision in the case of telecom operators) or another platform (personal computers in the case of Microsoft) does not guarantee you will earn the right to orchestrate a new platform (as Apple and Android have managed to do). Being a strong participant in an ecosystem (Samsung) does not guarantee that other players will support you to orchestrate it.

Today, the two companies orchestrating smartphone ecosystems are platform owners. They are able to control how business value is created and divided up at the intersection of the four business models. As we've seen, products and services on their own don't have enough scale, scope, or speed to solve customer problems. Platforms dominate currently because solution business models are expensive and not well enough established. However, as the mobile world further evolves from apps to bots, solution integrators could create new ecosystems in which they solve problems by negotiating across apps and services and generate enough value in such ecosystems to earn the right to orchestrate.

We are still learning about orchestration. Gone are the days when a company like Microsoft could dominate an ecosystem as it did with Windows because there were no alternatives. Apple is striving to manage a fine line between generating value from its own proprietary pieces and providing enough incentives for others to participate in its ecosystem. It does not welcome any hardware manufacturers to compete against its iPhone or Macintosh lines. However, it allows software applications from Microsoft, Google, and others to run on its computers, phones, and watches (and potentially on its televisions and other domains in the future). As you look at your own company, take stock of the ecosystems you are currently in or the ones you could become a part of. Who controls the various platforms? What value do you bring to the other players? Establishing your distinctive strengths will help you decide how to craft your first winning move when navigating across ecosystems.

DECIDE ON YOUR BEST ROLE

In any ecosystem, you have two strategic options, as we've seen. You can be the orchestrator of the ecosystem by coordinating the various building blocks to deliver seamless value to the customers. Or you can be a participant that delivers one or more of the important pieces that make the ecosystem distinctive. One is not intrinsically better than the other under all conditions. You have to choose the one that best matches your skills and capabilities. So how do you choose which one will bring you better value? Which one is the winning move?

The evolving transformation of the automotive ecosystem—with a greater role for electric drivetrains and cloud connectivity—helps us answer this question. Traditional ecosystems existed around individual car manufacturers, which controlled their supply chains and distribution channels. They integrated these value chains end to end, with different types of short- and long-term business contracts; in effect, each major car manufacturer orchestrated its own closed ecosystem. Recently, at least two digital giants have thus far shown interest in influencing the future of automobiles: Alphabet has the Android Auto ecosystem and Apple has CarPlay, both of them courting major manufacturers of branded cars and automotive electronics such as Pioneer and Kenwood.

If you are in one of those incumbent automakers, you are faced with a sequence of decisions as follows:

Should you participate in digital automotive
ecosystems proposed by Google and Apple?
The answer in most cases would be yes, given the growing importance
of software to provide key features that are increasingly controlled via
smartphone apps. The answer is no only if you intend to announce a
competing ecosystem in the near future. Even if you have ambitions to
architect a new ecosystem with another digital giant or by yourself, it is
advantageous to join the ecosystem now to learn and understand the
complexities and nuances of how digital features influence automobiles. You may find that their scope does not cover certain areas, which
may lead you to create your own focused ecosystem that might actually complement both Apple CarPlay and Android Auto.

Then, the next question is:

Should you participate in both ecosystems?
Here, you are making an important strategic choice. Your decision may
be heavily influenced by the fact that your customers in the automotive industry use either Android or Apple devices but not necessarily
both. This may lead you to support both ecosystems, as rejecting one
will simply alienate a part of your consumer base. More strategically,
if your rationale is not to preferentially align with one ecosystem, you
should support both. It would be different if one of these companies
had relatively few customers (as in the case of the Windows phone or
BlackBerry), but given the current global position of Apple and Android
in the automotive industry, you are most likely to participate in both
ecosystems. (Note that in other settings, that may not necessarily be
the best option.)

Now, for some deeper strategic thinking:

Should you explore a preferential
relationship in one ecosystem over another?
Here, if you're a carmaker, you must evaluate if working with Android
or Apple allows you to integrate more deeply with the automotive

dashboard or if it will lead you down a slippery slope in which your car becomes the hardware shell for software from these digital giants. It is not about the current level of functionality offered in Apple CarPlay or Android Auto but about how these software and apps could evolve to transform cars to computers on wheels connected to the cloud over the course of the next five years. This thought process leads to the next question:

Could you orchestrate mini-ecosystems
linked to both Apple and Android?
Here, you could look at a broader set of relevant actors, such as telecom operators like AT&T, which has announced cellular connectivity to cars; subsystem manufacturers such as Bosch, Dana Incorporated, and Continental tires in Germany, which has an experiment with Google to optimize the performance of tires; cloud computing providers such as Salesforce.com, which has demonstrated ways to connect Toyota cars to the cloud; and Ericsson, which has worked with Volvo to build and operate Volvo Cloud to connect cars and collect data with other cars, the company, and local municipalities in Sweden and Norway.

I said that that you had two options: to orchestrate or to participate. But the truth is that these roles represent two extremes in ecosystems. Since ecosystems are not static, the roles can also vary. Participants who want to gain a slightly higher share of revenue and profits can work with other participants to create mini-ecosystems within the larger one. Ultimately, Apple and Android's ability to orchestrate their automotive ecosystems depends on the support of the product and service companies that subscribe to their platforms and choose one over the other. You may not be able to call all the shots, but your actions *are* important to the success of the orchestrator and their ecosystem. As an industry incumbent in the automotive industry, you do not have the option to stay on the sidelines. You have to step up to the digital business arena and align with one or more players that are already aiming to orchestrate. Or you may aim to be the orchestrator yourself by acquiring the necessary new capabilities and assembling the necessary partnerships; this is not a tactical decision but one that could well define your future course of evolution. Remember, the digital giants may have an edge in orchestrating in the early stages of digital transformation, because they can draw on their established scale and scope to lead, but incumbents can regain and seize the advantage.

The most strategic question facing you as an incumbent is this: How could you work with digital giants and other entrepreneurs to ensure that your role—either as a participant or as an orchestrator—enables you to get a fair share of the economic value as the automotive sector transforms?

Learning from Walmart

Now, let's look at a different ecosystem: retail payments, which are transitioning from magnetic and chip-and-PIN cards towards software apps inside smartphones. The transformation cannot happen overnight, because of the legacy infrastructure involving terminals and card readers, but digital payments are clearly on their way. Here again Apple and Google emerge as potential orchestrators, with Apple Pay and Android Pay. Just as in the mobile broadband ecosystem, Samsung has ambitions to be more central; it has its own Samsung Pay. There are, of course, also the traditional players, such as issuing banks (Chase, Bank of America, Barclays), card networks (American Express, Visa, MasterCard); acquirers and processors (Chase, Cielo, Citi); payment gateways (PayPal, Stripe, Klarna); and a wide array of merchant service providers. How could the digital payment systems evolve over the next decade and who could orchestrate them?

Let's try to answer the orchestration versus participation question through an incumbent company such as Walmart, using the same set of questions as before.

I. Should Walmart participate in the retail payment ecosystems proposed by Google and Apple?

2. Should Walmart participate in both ecosystems?
3. Does Walmart have unique skills and capabilities to explore a preferential relationship in one ecosystem over another?
4. Are there opportunities to further leverage Walmart's own digital capabilities to orchestrate mini-ecosystems that could link to both Apple and Android?

When Apple introduced Apple Pay in 2012, Walmart would have answered no to all four questions, mainly because it wanted to orchestrate digital payments, at least around digital commerce, online and in stores. Long before Apple and Google announced their intent to move into retail payments, Walmart had pulled together a consortium of merchants in an ecosystem around a vision to create CurrentC. It was positioning to orchestrate its own payment ecosystem and had strong backing from retailers such as 7-Eleven, Target, Best Buy, and Lowe's. So Walmart's initial response to Apple's payment ecosystem was to reject it in favor of continuing to show commitment to the Merchant Consumer Exchange (MCX)[4] and its projected CurrentC payment system. As a would-be orchestrator, it persuaded the member community to reject Apple Pay; it wanted to prevent its participants from defecting to this competitor's ecosystem.

The participants, however, were divided about what to do. Some merchants (e.g., CVS, Target) decided to shut out Apple Pay in their stores and stay true to the original intent of the MCX, but others (e.g., Rite Aid) wondered whether to continue to participate in this initiative exclusively or to renegotiate the contract to explore working with Apple Pay (and whatever else might come later). This dilemma was a classic reaction to collision at the core, with new players offering newer options. With more options to participate in than before, some merchants wanted to choose the system with the broadest scale and customer appeal, ease of use, and reliability (with better security features). They were no longer interested in CurrentC, which appeared to be an inferior system after the introduction of Apple Pay and other variants.

In late 2015, Walmart still continued to support the collaborative initiative, but concerned that CurrentC might not gain the necessary support of other retailers, it introduced its own proprietary Walmart Pay[5] as a competitor to Apple Pay. Walmart's standalone currency may gain traction within its stores and with its customers, who could be lured with attractive offers (as was done with store loyalty cards in the late 1990s), but for the company to credibly orchestrate its own retail payment ecosystem, it needed more participants.

The company leads retailers in terms of the number of physical stores and number of loyal customers who shop on a weekly and monthly basis, but that scale does not logically translate to orchestrating digital payments. In mid-2016, MCX and the CurrentC system have been all but shuttered. Walmart still has aspirations to orchestrate, but so far, it's only leading its own closed payment ecosystem with very little participation from other retailers.

LEARN ACROSS DIFFERENT ECOSYSTEMS

What do the examples from the automotive industry and digital payments show us? Basically, if you are a small or medium-sized incumbent, you may not be able to effectively orchestrate ecosystems that compete against the scale and scope of the digital giants, which can extend their existing ecosystems across industry boundaries and leverage their ecosystem advantage. You should participate, perhaps in all of the ecosystems, perhaps preferentially with one or more. As you gain distinctive proficiency, you should step up to orchestrate focused ecosystems as a way to learn and expand your networks. If you are a larger incumbent, you can't assume that your scale alone in your traditional industry is going to give you an advantage in the digital realm. If you already have strong digital capabilities and ecosystems that you can leverage, consider positioning yourself to be the orchestrator of your own ecosystem by bringing your history of relationships and goodwill in the industry. Be aware, though, that your would-be partners will be evaluating your ecosystem against the digital giants' using a wide variety of criteria. Your history of past relationships may not outweigh the intrinsically superior digital functionality, such as privacy, security, cloud connectivity, and data analytics that may be available in competing ecosystems.

The digital giants will jockey to orchestrate ecosystems in every industry, so it's important to observe and analyze what they are doing and prepare your move. At the same time, it's wrong to assume that they have an intrinsic advantage in all sectors, under all conditions. Apple, for instance, has not been able to orchestrate a new television experience, as major television channels and cable operators have not (yet) agreed to Apple's conditions. Essentially, Apple has not yet earned the right to orchestrate the television ecosystem in the same way it has in music with iTunes and iPod. The same is true in the automobile sector and in industrial ecosystems such as food and agriculture, chemicals and fertilizers, buildings, electricity, transportation, and other sectors in which there are still no clear orchestrators. Companies such as GE, Samsung, Bosch, ABB, Siemens, and others have all revealed their intent to orchestrate in these areas.

Again, the scope of ecosystems is dynamic and has multiple platforms, products, services, and solutions. When traditional and digital technologies collide, traditional industry roles matter less, historical ways to earn revenue and profits go away, past characterizations of strengths and advantage do not hold sway, and some activities simply get eliminated and disappear. Will emerging ecosystems be defined around what the digital giants see as the business architecture or will traditional players get to jointly (or even individually) define it? Those are the essential questions now facing every digital transformation setting—including your own.

DO YOU HAVE WHAT IT TAKES TO ORCHESTRATE?

It's important to recognize that the role of the orchestrator is not static (and anointed for long durations), but there's constant struggle to ensure that the best orchestrator is chosen to maximize the value of the ecosystem. The retail payment ecosystem and Walmart's struggle in it is a typical case. A successful orchestrator ensures the following six characteristics:

Vision. An orchestrator articulates a compelling vision to solve important customer (or societal) problems. Uber's aim to help consumers with transportation and mobility in cities is one; and Google's vision to accelerate smartphone adoption at affordable price points is another. Might Apple's vision be compelling enough to orchestrate in the personal health ecosystem? Will Alphabet truly pull together a network of ecosystems with overlapping visions to transform health care? Can GE orchestrate industrial ecosystems in different arenas to achieve new levels of productivity? Who will develop a compelling vision for the blockchain that appeals to a variety of different industries? Time will tell. In a nonbusiness setting, the Bill & Melinda Gates Foundation articulates a compelling vision to attract other philanthropists such as Mark Zuckerberg and Warren Buffett to join forces. The vision needs to be compelling to draw other participants to it; it should be sharp and differentiated for the key participants to not want to orchestrate it themselves.

Distinctiveness. An orchestrator follows up the vision as articulated with a unique critical capability required by participants and valued by customers. As Uber competes with Lyft, Didi, Gett, and other transport-on-demand services, what distinguishes it from the competition? Why should the freelance driver prefer to participate in this ecosystem versus another? Why would riders prefer one to another? The economic principle of network effects is one reason, but over time, as drivers link to Uber and its competitors and as riders install more than one app, the switching cost for both sides is zero and other areas of distinctiveness must be probed. Walmart's inability to orchestrate the retail payment ecosystem is partly because of its lack of distinctiveness relative to Apple, Samsung, Android, PayPal, and others. This distinction needs to be articulated for two sets of players in every ecosystem: the participants that provide its strength and the customers that provide the money.

Complementarity. An orchestrator attracts complementary partners that enhance the value of the ecosystem. The ecosystems grow through a virtuous cycle involving acceptance by customers (researchers label these "direct network effects") and participants providing support through the four business models discussed earlier (researchers label these "indirect network effects"). The more diverse the complementary participants that are attracted to the orchestrator, the more vibrant and useful the ecosystem. With this in mind, Uber links to OpenTable, MasterCard, Hilton, and others with deep app integration to make it easier for customers and participating companies. We saw it with Windows and the wide variety of software applications that ran on the os we see it now with mobile os; and we will see it with successful payment ecosystems, with smart home ecosystems, with digital health ecosystems, and so on. Ultimately, the orchestrator pulls together a *system of network effects*[6] across multiple different levels. An ecosystem grows in power and importance with *systemwide network effects* that "take hold when adjacent parts of an overall system are built out—e.g., smartphones, wearables, sensor networks, new physical layers, blockchain etc. Network effects at these layers are incredibly powerful as they effectively unlock compounded value from previous layers."[7] So the digital giants, to the extent that they can tap into such systemwide network effects, are in a stronger position to orchestrate than incumbents. It simply means that incumbents must step up to take advantage of newer layers of digital infrastructure being built now with the Internet of Things, the blockchain, and the cloud.

Respect. An orchestrator earns the right from other participants. In some ecosystems, participants may provide support because they do not have any viable, credible options, but they are always looking for the first reason to break free and support another plausible one. Such participants do not have deep respect and are only showing obligatory (but not deeply substantive) support and will defect when given the first chance, especially with low switching costs. Orchestration with such weak support is like quicksand, without foundation. A company's ability to continue to play the orchestrator role depends on the respect shown by the participants in its ecosystem during periods of uncertainty and technological shifts. An orchestrator that wields "soft power"[8] (the power to co-opt) rather than "hard power" (the power to coerce) is more likely to gain respect.

Governance. A successful orchestrator manages the role conflicts that invariably arise within and across the ecosystem by being fair and transparent about the way they are handled. For example, IBM, as a solutions integrator, must delicately balance the use of first-party components (those produced by IBM) with third-party parts that best deliver value for its clients. Amazon must delicately balance how not to favor its own products (e.g., Kindle) over competing products (Apple iPad) in its stores. And app stores orchestrated by Apple and Google must delicately balance the way they handle third-party apps that could be seen to conflict with their own or some preferred apps— such as maps, payment, photos, and music.[9] There will be inevitable tensions about an orchestrator's perceived preference for some participants over others or for their own products and services over others. If so, are these openly stated and consistently applied? Under what conditions are exceptions made and how well communicated are they? Trust is significantly enhanced when the enacted rules of governance match what the orchestrator has proclaimed up front.

Dynamics. An orchestrator proactively adapts the vision and interconnects with other ecosystems (as necessary). An ecosystem's participants expect the orchestrator to be in tune with the direction and pace of change and help them along. Mobile apps have evolved with greater functionality of hardware, including screen size, form factors (e.g., phone versus watch versus tablet), and interface (touch versus voice). So participants expect the orchestrator to keep abreast (if not lead) with technological developments so that they don't feel trapped by lagging functionality. In the case of video games, for example, the console players such as Nintendo and Microsoft (Xbox) have helped the game developers across successive generations of game console architecture with tools and guides to port their existing games. Not keeping up with the dynamics is one sure way for orchestrators to lose the support of their participants.

HOW TO DESIGN YOUR FIRST WINNING MOVE

In the digital business world, every company is embedded in multiple ecosystems that can be understood using the four archetypical business models—product, service, platform, and solution. It's important to recognize you could well be a member of more than one ecosystem, which means that it is possible for you to be the orchestrator in one (or more) and a participant in others.

The six principles give you a working guideline to decide where you could credibly step up to orchestrate and when to provide support to someone else to be the orchestrator. With the recent frenzy around digital business platforms—Airbnb, Uber, YouTube, PayPal, Alibaba, and others—nearly everyone wants to be on the "platform bandwagon." Unfortunately, not everyone's capable of architecting the platform and orchestrating the ecosystem around their vision with distinctiveness and credibility. So it is neither necessary nor advisable for you to be orchestrating every ecosystem. An important part of the first winning move, when you're navigating ecosystems, is to decide the areas in which you want to orchestrate and those in which you want to participate, and adapt these choices as conditions change.

Three steps help you design the first winning move.

Define your relevant set of ecosystems. No matter what phase of transition you are in, you are not located in just one neatly defined ecosystem but in a set of interconnected ecosystems. For example, if you are an incumbent automaker, you could be in an ecosystem for parts and subsystems (with other automakers), another for telematics-related services (with telecom operators and cloud service providers), another for mobility and applications (with digital giants working to define how the apps work across devices without sacrificing safety), and another for transportation services (with all three players working across multiple modes), and so on. As the car industry further digitizes, additional ecosystems will emerge (for example, an ecosystem for hybrid powertrain or for automobile batteries) and the relative importance of the existing ecosystems will undoubtedly change. Automakers may not have prioritized battery ecosystems in the early 2000s, but by 2014, when Tesla announced its gigafactory in Nevada with Panasonic as a lead partner,[10] this ecosystem grew in importance. By 2020, Tesla may be the orchestrator of this ecosystem, if it produces nearly 50 percent of the planned production. Uber may be the orchestrator today, since the automakers were not thinking broadly about their scope, but with initiatives from GM with Maven, Ford's mobility experiments, and Daimler's Moovel (and others), we should not rule out possible orchestration of urban mobility by automakers.

If you are a pharmaceutical incumbent like Merck or Novartis, you are beginning to see early moves in digital health ecosystems in areas such as health IT platforms, electronic and personalized health records, wearable technologies and health and fitness apps, and advanced analytics. These ecosystems today are characterized by fluid boundaries with all three types of players staking claims to orchestrate one or more ecosystems. They may appear ill formed at the moment, but very soon, these ecosystems will grow and become central to the very process of creating and capturing value in health care. As that happens, the orchestrators will become clear. For example, wearable devices for health and fitness using primitive sensors may still be experiments at the edge, but by 2020, those sensors might well be medical grade and the devices capable of a variety of diagnostics. The data gleaned from such sensors could define new ecosystems involving not only health care and fitness but also insurance and other related domains.

A simple approach that begins with how customers perceive value is useful. Start with a list of ecosystems that are central for delivering value to customers, because as I have highlighted in this book, value delivery to customers is through a portfolio of capabilities assembled through a network of relationships. For each ecosystem, identify who's best positioned to orchestrate it now: Is it your company or someone else? Analyzing how you and the others stack up against the six criteria discussed above helps you to decide your role.

Decide on your role in each ecosystem. As we discussed earlier in this chapter, you have two choices in each ecosystem: to orchestrate or to participate. It is no longer a good strategy to try to orchestrate in every ecosystem (even Apple and Alphabet don't do this), so think hard about where you want to orchestrate and where you want to participate. GE has taken the lead with the Industrial Internet, but there are many other global companies in the midst of their own digital transformation. ABB, the Swiss industrial company, has ambitions to differentiate its offering with software-based innovations in areas such as mining, energy, and power grid infrastructure, and has entered into partnership with Salesforce, Statoil, Bosch, and Ericsson.[11] My advice to such industrial companies that are facing digital influences is that you will find yourself in many different but interlinked ecosystems. So plot your list of ecosystems on a grid: Draw a square and divide it into four quadrants. Label the vertical axis with the strategic importance of the ecosystem (high or low) and the horizontal axis with your choice (participation or orchestration). Then write the names of the ecosystems you're involved in in the appropriate box. Are you satisfied with your positions, and if not, what needs to be done as you further evolve along your digital transformation journey?

Compare your profile against those of your immediate competitors to discern where your advantage lies. Your digital business strategy is not defined by the technologies you have experimented with or licensed but by clearly articulating your roles in complex and dynamic ecosystems formed at the intersection of your traditional industries and digital frontiers. Remember that these roles of participation and orchestration could well change from one phase to another; I discuss the dynamics next.

Examine the dynamics of these ecosystems. This is the key part of strategic thinking. With the speed of digitization, you will find yourself in the midst of new sets of ecosystems—some extensions of the current ones that you defined in step 1, others entirely new because of creative innovations by the three sets of players. Think about how new digital technologies, such as 3D printing, the blockchain, and collaborative robotics, might introduce new digital business ecosystems in your industry. For example, Siemens is considering using blockchain technology, essentially a digital public ledger, to replace the tokens used in washing machines in apartment complexes or to keep track of surplus solar power sold from one neighbor to others on the same street. If you are Siemens, does your existing role as an orchestrator or a participant in the blockchain ecosystem change? Referring to the grid of ecosystems you created above, under what conditions would you proactively shift your role? As you reflect on the six criteria discussed earlier, what new capabilities might you need to become an effective orchestrator?

UNDERSTANDING DIGITAL BUSINESS ecosystems and whether to participate or position yourself to orchestrate is an important winning move, because it allows you to know where to apply your energy and how to make the most of your resources *across* relevant ecosystems. It also allows you to clearly identify who are your competitors and who are your potential allies—and when—both inside and outside of your existing ecosystems. That's a key piece of information—and the subject of the next chapter—that will help you navigate effectively within ecosystems.

CHAPTER 7

COLLABORATE TO CO-CREATE NEW CAPABILITIES

MOST PEOPLE THINK about business as a competitive environment. We compete for market share, we compete for employees, we compete to earn more revenue than our rivals. But as we know, there are also times when we have to cooperate. We cooperate with our suppliers and distributors, we cooperate with our customers, we cooperate with lawmakers and regulators. And we love to categorize things as being in one camp or another. Digital business, with its emphasis on networks and working in ecosystems, blurs that distinction. Or as Ray Noorda, former CEO of Novell, stated when he coined the term "coopetition," "You have to compete and cooperate at the same time."[1] When it comes to working within ecosystems, this is your winning move. Work not only with your allies, but seek out your competitors so that you can learn from them and collaborate to co-create new offerings and develop new capabilities.

WHEN AND WHY COOPETITION IS KEY

In 2016, most observers probably see Google and Apple as competitors, especially when it comes to mobile phones. But that wasn't always the case (and it isn't entirely true). When Apple CEO Steve Jobs launched the iPhone in 2007, he had invited his counterpart at Google, Eric Schmidt, onstage to show that Google and Apple were partners in this innovation. Schmidt was a member of Apple's board of directors. Apple iPhone had preinstalled Google Maps, and Google was the default search engine. In the Jobsian vision, Apple—working alongside Google as a preferred partner—would disrupt and dominate telecom, just as Microsoft had done working with Intel on personal computers just two decades before.

Apple saw its relationship with Google as a preferential one, with complementary (non-overlapping) capabilities that would endure for some time. In contrast, its relationship with AT&T, the preferred launch telecom partner with exclusive rights to distribute iPhone in the United States for two years, had an expiry date. Apple would naturally work with other telecom carriers to distribute future versions of the iPhone to achieve global scale, but no other map apps would be preinstalled on the iPhone. Apple had not created its own map app to compete against Google by then and had no intention of doing so. The two companies were collaborating to co-create value with smartphones; there was no competitive friction or combative tension between them.

By 2009, two years after the launch of the first version of the iPhone, Google launched the Android operating system (OS) to compete with Apple's iOS. Unlike Apple, which made both hardware and software OSs for phones in a tightly integrated fashion, Google chose to make Android OS available to a network of hardware manufacturers to design and manufacture Android-compatible devices. Apple and Google had gone from being partners and co-creators of the iPhone smartphone to being competitors. Although Google's Android OS did not directly earn revenue, the Android ecosystem competes head on against Apple.

Steve Jobs exploded and declared that he was willing to go to thermonuclear war to get rid of Android.[2] He squarely blamed his erstwhile partner for the change of position in their relationship: "We did not enter the search business. They [Google] entered the phone business.

Make no mistake: Google wants to kill the iPhone. We won't let them."[3] Although Google had bought Android as early as 2005, Jobs believed there would be no direct competition between his company and Google.

Why did Google develop Android to compete against its trusted partner? In my view, the compelling business rationale was to protect its core business. The digital world was evolving from desktop to mobile, which threatened Google's core advertising business. If Google could (and did) not adapt its advertising business to mobile, it would miss the next turn of the technology revolution. So Google's instinct was to go where the search was going and bring its advertising capabilities to the mobile world. In doing so, its cooperative relationship with Apple turned adversarial and competitive. Apple developed its own maps app but has retained Google as the preferred search engine (since it does not have its own search engine), with Google reportedly paying Apple about $1 billion for the privilege.[4]

Today, out of necessity, they are in fact cooperative, competitive, and coopetitive, depending on the time and the industry. As Ray Noorda recognized in the early 1990s, "We are moving toward a survival environment, and the direction will be set by those who have become the strongest."[5] That strength comes from co-creating value for consumers. To satisfy consumer demands for interoperability across hardware, software applications, and services, major Google apps are on Apple's app store and Apple apps are on the Google Play store. They have similar points of view on security but differ on privacy and the legitimate use of consumer information for monetization purposes. They compete for the same talent pool in Silicon Valley; they compete in media and entertainment (Google with YouTube, Chromecast, and Google Video, and Apple with iTunes and Apple TV); in browsers (Chrome versus Safari); in assistants (Siri versus Assistant); in photos, storage, and mail (Gmail versus Apple Mail). Google is strong on mobile ads; Apple has retreated. As we saw in Chapter 6, they are the two major—competing—orchestrators on the mobile web, and they are locked in other areas such as the automobile industry and health care.

The Google-Apple interconnection over the last decade provides useful pointers about the fundamentals of business relationships and the new frontier of co-creating value in digital ecosystems. The relationships among companies in evolving ecosystems are fluid and dynamic, as consumers want change and companies alter their set of capabilities to go after new pockets of value. As a result, no two companies are likely to be purely competitive or purely cooperative across time because their capabilities are not neatly demarcated but complexly interdependent. And every company enters into many types of relationships: some focused on the current time period for efficiency and value capture, others directed at innovations that might need a longer lead time to generate value. Managing such relationships to continually fine-tune your capabilities is key.

ENLARGE THE VALUE PIE
BEFORE YOU DIVIDE IT UP

The idea of coopetition highlights how strategies for value creation and capture in the digital world differ from the classic principles refined in the industrial age. Pure competition is about *dividing up* the existing value pie; companies use their set of capabilities to win a greater share of the value. Cooperation is about *enlarging and expanding* the value pie by pooling the capabilities of several companies for the short term and also for the longer term. In contrast, coopetition is about both expanding the pie *and* ensuring that you get a fair share of the value.

So at any given time, how do you determine who to cooperate with, who to compete against, and who to both cooperate with and compete against? In 1996, Adam Brandenburger and Barry Nalebuff wrote an influential book titled Co-opetition: A Revolutionary Mindset that Combines Competition and Cooperation to get at just that question.[6] They used game theory to develop a framework to introduce the role of complementor, an addition to the familiar business system of company, suppliers, buyers, and competitors. They determined that:

1. A player is your complementor if customers value your product *more* when they have the other player's product than when they have your product alone.
2. A player is your competitor if customers value your product *less* when they have the other player's product than when they have your product alone.

So many of the platforms in digital business settings are comple-mentors to the products and services that run on them. For example, the presence of Windows 10 enhances the value of a range of software applications that have been specifically designed to run on it and take advantage of its functionality. A set of hardware devices views Windows 10 as a complementary platform as well. However, since platforms vie against other platforms to attract products and services that make their platform more attractive than others, they are also mutually competitive. So Windows 10 competes against Apple (MacOS and iOS) and Google (Chrome and Android) in terms of their hardware devices and software applications.

Since the same player can be both competitive and cooperative (as we saw with Google and Apple), these relationships need to be managed differently than if the complementor is not a competitor. Moreover, the relative dynamics of competition versus cooperation change over time, especially as the digital architecture of business evolves across the three phases of the transformation.

You, as a leader in an industry incumbent, are very familiar with long-term supply contracts, multiyear manufacturing deals, or exclusive marketing and distribution agreements. You have long-established processes and best practices. Today, your business ecosystems may look neatly defined around your supply chains and your distribution channels, with very little overlap in areas that are central to value creation and capture. The main suppliers and distributors up and down the value chains are mostly with you or with your competitors. You also understand intuitively the power of coopetition in the business realm of hardware and software, because of the inherent interoperability and sharing of data through applications programming interfaces (APIS)—tools and protocols for design and use across different systems—and connectivity. You have also seen the results of such coopetition in the ways your information technology operations function and deliver services to your employees. If you are involved in the management of IT within your organization, you have seen how coopetition within technology vendors—Microsoft, Salesforce, Box, VMware, IBM, Cisco, Amazon, and others—has helped you migrate your IT operations to the cloud.

In essence, then, an ecosystem working with complementors expands and enlarges the value pie for everyone because it allows all of the businesses on a platform to take advantage of scale, scope, and speed to reach more consumers across a wider range of industries more quickly and with more targeted offerings. The sensors and software and service delivery inherent in digital technologies pull together data from multiple different areas, interconnecting incumbents with tech entrepreneurs and digital giants simultaneously as competitors and collaborators. With increasing digitization, you will see more coopetitive relationships, and your ability to recognize the shifts and respond to them becomes central for capturing your fair share of the value pie and developing the necessary capabilities to ensure your future success.

RECOGNIZE HOW DIGITIZATION
INTENSIFIES COOPETITION

Coopetition is central when traditional industries digitize because a) value is co-created with multiple companies across these traditional industries, and b) digital industries combine in ways to deliver new services to customers and capture value. Every company must coordinate the design and delivery of its products and/or services with platforms and solutions companies in ways to deliver the value that customers want. This is the essence of what's called "capability co-creation."

In the pre-digital phase, you and your industry incumbents will traditionally have treated the digital giants as suppliers and not as competitors. However, in phase I of transformation, when experiments are taking place at the edge of your industry, you might begin to see the digital giants and tech entrepreneurs as collaborators whose digital knowledge can help you sharpen your own. By phase 2, many of these same companies may have become competitors to you or to other incumbents, as their incursions into your industry collide with and disrupt your traditional ways of doing business. They may be solving your customers' problems in ways that you are not. As you're reinventing your business by phase 3, however, you and the tech entrepreneurs and digital giants are more likely to be in competitive, collaborative, and coopetitive relationships within ecosystems. You will have accepted that *you will compete against and cooperate with other companies* in dynamic and complex ecosystems. That means you will invariably be co-creating capabilities with companies that may be competing against you in some parts of the digital business system.

Learning from the digital giants

Let's do a simple exercise to map the relationships among some of the digital giants—say, Amazon, Apple, Facebook, Google, IBM, Microsoft, Netflix, and Samsung. If I asked you to pair them up and describe the relationships between the pairs as either (–) competitive or (+) cooperative, you might say:

- Apple and Google: – (operating systems, browser, email, etc.)
- Google and Facebook: – (mobile ads)
- Apple and Microsoft: – (desktop, laptop, and mobile devices)
- Amazon and Netflix: – (video streaming)
- IBM and Microsoft, Facebook and Google: – (artificial intelligence and cognitive computing)
- IBM and Apple: + (enterprise applications running on Apple OS)

You do not need to actually draw all the possible links to know that these companies are mostly competitive.

Where it gets tricky and subtle is if I asked you to add the third type of relationship: (+/–) coopetitive.

- IBM and Apple: We said they are + now, but what about IBM's Watson and Apple's Siri (both pushing next-generation experience through cognitive computing)?
- Apple and Samsung: – (mobile devices) but + (electronic components for the phones; indeed, Apple is the largest customer for Samsung's semiconductors division)
- Netflix and Amazon: – (video, albeit with different content and business models) but + video streaming (Netflix relies on Amazon Web Services)

When we delve deeper and complete this mapping exercise for the digital giants, in effect, every relationship is coopetitive (+/–) to a varying degree. These coopetitive relationships are multilayered: competition occurs in one layer and cooperation happens in another. And these are just a few examples. When you look at application programming interfaces, or APIs, you see that all of these digital giants are densely interconnected, with deep plumbing that ensures the interoperability of their services for consumers and, increasingly, for enterprises, too. Similarly, the Internet of Things requires cooperation relating to security, privacy, and identity. Over time, the networks of digital relationships become more coopetitive for ecosystems to gain scale, scope, and speed.

We have not even begun to map how these relationships extend to your industry across the three phases of digitization—then we have to bring the tech entrepreneurs and industry incumbents into the picture. This exercise is something I encourage you to do, but let me help set up your thinking about such interactions using four different zones.

CO-CREATE CAPABILITIES WITHIN ECOSYSTEMS

By now, it's clear that digital business calls for multiple types of relationships among the three types of players. As you and your incumbents work with digital giants and tech entrepreneurs to create the new foundations for value creation and capture in the digital era, you need to ask two questions. First, what's my level of importance to my partners? And second, what's my partner's level of importance to me? (For the purpose of this discussion, by partners, I mean all three types of players; you may want to separate them out later to develop deeper insights when selecting one specific company over another.) You may be competitive, cooperative, and coopetitive, but it is important to understand the relative importance that sets up the basis for interactions within ecosystems. If we plotted the possible responses on a grid, we'd get the following four zones (see Figure 4):

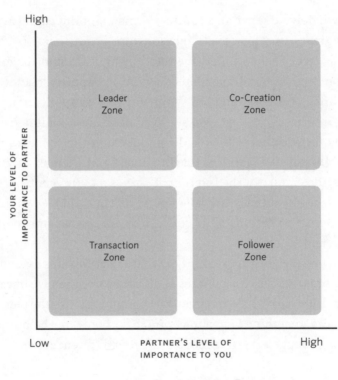

FIGURE 4 Four Interaction Zones

Transaction zone. If both you and your partners see little value in interacting either to cooperate or to compete, your level of coopetition and co-creation will also be very low. In this zone, which is typical of phase 1, you could be experimenting on your own and the other players could be working on their own models of disruption. If there are any transactions between you and others, they are likely to be based on standard contracts as buyers and sellers. That is, you may be focused on digitization trends and options, but you're thinking about what you could do by yourself with resources that could be acquired rather than through connecting with others. In this zone, there are really no specialized relationships and you really are not (and should not) be spending much time thinking about coopetition. This zone becomes important again once you've established your priorities, as you may shift some of your active relationships here while you give more attention to others.

Leader zone. You as an incumbent have recognized your company's need to transform its business model and are working with others who are currently less invested in the relationship; you are in this zone as long as you are working from a position of strength. You are confidently defending your core strengths while adapting to digital ways of working. You recognize the important assets and capabilities that the other players bring to the marketplace and involve them as necessary to complement what you are doing without allowing them to be deeply involved in your transformation. You may enter into a more preferential arrangement than standard contracts, but you are careful to protect your own advantages during this transformation. A great example is Ford, which initially worked with Microsoft to create SYNC but later decided that its digital services road map could be better designed with BlackBerry's QNX operating system. Daimler, GM, Toyota, and others have pursued this line of coopetition until now but could well need to shift to a different zone as their digital transformation unfolds further.

You could be in this zone during any of the three phases of transformation because you have chosen your partners carefully from a position of strength. If a digital partner does not fit in with your selection criteria and frames of agreement, you might select a tech entrepreneur to work with you in this zone. Or you may involve industry incumbents to jointly explore new directions. In essence, you are leading your digital transformation shifts, and the other players are the supporting cast. As you understand the speed of digitization, you may want to assess how and when you could locate some of the key initiatives in this zone, even if you started off in the follower zone.

Follower zone. This zone is the polar contrast to the leader zone. The other players are in the position of directing the transformation agenda, and you are more or less following the course of action defined by them. Speaking about the digital giants creating mobile software in the automotive industry, Fiat Chrysler CEO Sergio Marchionne called them "disruptive interlopers."[7] Beyond the auto sector, in your own setting, you may realize your lack of necessary competencies as your industry digitizes and have no plausible option but to work with some of these disruptive interlopers, at least in the immediate term. Earlier, we saw that Honeywell had to deal with this issue when designing its Lyric thermostat, as every major device manufacturer will as they face the future smart home. In many industries, digital players have stepped up to lead the agenda, and you realize the inherent coopetitive tensions but are pulled into situations where there are no other attractive options for the short term. You may enter into a specialized relationship that gives you follow-on options, but in most cases, you are working together so that you learn quickly and understand the likely scale, scope, and speed of digitization for your industry and business.

Your challenge, then, is to think through not just the first stage of coopetition but also your follow-on decisions to ensure that this digital transformation journey does not weaken your competitive position in the next time frame. TAG Heuer currently faces this situation. When Apple launched its Apple Watch, TAG Heuer had to decide whether to develop its own smartwatch, and if so, whether to do it on its own or with a digital partner. The CEO, Jean-Claude Biver, chose to go with Google and Intel[8] to develop the Connected Watch, which has been a successful addition to its brand portfolio. He admits that there was no other way to develop a compelling connected watch without these two partners.

The key question for TAG Heuer is about how it will manage its future relationship with its digital partners: What happens when another watchmaker approaches Google and Intel? Will these giants be content to be work exclusively with TAG Heuer, or will they logically expect to expand their scope? It is a fair guess that these two digital players will want a wide range of branded watches on the Google-Intel platform, just as we've seen with other computing devices over the last several decades. So TAG Heuer is going into this digital avenue expecting that, even in a crowded arena of other smartwatches on the Google-Intel platform, its brand name and design will keep its smart-watch distinctive. This early move gives the company the insights needed to develop the necessary digital capabilities in-house or work with others, but by shifting to the leader zone.

Chances are that your current approaches to co-creating capabilities fall within one of these three zones; some may be in the transaction zone, if they are not of much importance. Some may be in the leader zone, if you have proactively stepped up to the digital shifts, perhaps during the collision phase as you seek to morph your business logic, as GE has done with the Industrial Internet. And some may be in the follower zone, if you are seeking to coexist (as with TAG Heuer). The fourth zone becomes important as you focus squarely on the future, especially during phase 3, reinvention at the root. Let's look at that zone now.

Co-creation zone. When you and your partners are working together to co-create value that neither of you could have done alone, then you are both highly important to each other. Co-creating through coopetition is both important and central to reinventing your business in phase 3. Lenovo's relationship with Microsoft is an interesting example. Satya Nadella, CEO of Microsoft, has revived the company through its Windows 10 ecosystem. However, unlike when former CEO Bill Gates orchestrated the Windows 95 ecosystem, Microsoft has done it by orchestrating new ecosystems with partners that believe in the future of the Windows 10 operating system.

Since 2005, Lenovo has been creating hardware devices to support the Windows platform; in fact, after the launch of Lenovo's Yoga personal computer in October 2015, Joe Belfiore, Microsoft's head of Windows 10, blogged: "We worked closely behind the scenes with Lenovo to ensure its new products really brought to life the best of Windows 10."[9] Clearly, Microsoft wanted to help Lenovo create the best device so that Microsoft could reach its goal of 1 billion devices—from personal computers to handhelds and wearables—running on their latest operating system. And Lenovo wanted to develop a PC that worked seamlessly on the coveted Windows 10 platform.

So far, so good: Microsoft and Lenovo are co-creating value for each other. (Every one of Microsoft's major hardware partners could be in this zone.) But Microsoft has its own line of Surface-branded products that competes against Lenovo. When Microsoft asked Lenovo to resell Surface in addition to selling Lenovo-branded products, Lenovo COO Gianfranco Lanci explained: "I said no to resell their product … [Microsoft] asked me more than one year ago, and I said no I don't see any reason why I should sell a product from within brackets, competition."[10] He further remarked that his company viewed Microsoft as a "partner on certain things" and a "competitor" on others. That is the essence of coopetition, and it highlights the essential tension every company faces managing relationships in this zone.

Every coopetitive relationship is different because all of them are multilayered, but what all of them share is a common understanding of what each party brings to the relationship and what their motives are. As we've seen, Apple and Samsung compete with mobile devices but cooperate on components. Apple is Samsung's biggest external customer, delivering flash storage, application memory, and display screens. It also manufactures the A9 chips (based on Apple's design), because it does so with better precision and security than anyone else. Apple gets a superior product that safeguards its intellectual property, and this reinforces Samsung's reputation for quality and reliability. Similarly, Netflix and Amazon Web Services compete on video on demand at the same time that they cooperate on video streaming. By working with Amazon Web Services, Netflix gets thousands of servers and terabytes of storage within minutes, which allows it to deliver the best available service to its subscribers. At the same time, Amazon Web Services' reputation as the industry leader in video streaming is reinforced. Amazon must ensure that it provides the best possible service to Netflix, or it will lose credibility. Samsung's device and components business must maintain and safeguard Apple's intellectual property to retain its credibility. And Microsoft must work with its hardware partners to make sure that their products bring out the best of Windows 10, even if they beat Surface in terms of design, features, and performance.

Every coopetitive relationship also has different time horizons. Of course, you will not be able to precisely specify how each of the different types of relationships that you find yourself in will evolve, as we saw with Apple and Google over the course of a decade. The Amazon-Netflix relationship seems to be in a relatively steady state (with Netflix explicitly recognizing the trade-offs in either creating its own video-streaming system or working with the second best in the industry). Apple and Samsung seem to be on less solid ground for their future relationship, with Apple looking but unable to find credible alternatives. Time will tell how much Apple invests to create another credible option to Samsung as its supplier. Microsoft and Lenovo (in fact, all its hardware partners) reveal the tension in managing first-party (own brand) versus third-party (supporting partner's brands) so as to maximize the value in coopetition. In this zone, both companies need to be more adaptive and master capabilities that resemble less joint planning and more dynamic improvisation[11] with mutual respect and trust.

Learning from Sir Martin Sorrell at WPP

Sir Martin Sorrell is the chief executive of WPP, a leader in marketing communications. In 2009, as Google was beginning to tread on his traditional turf with digital ads, Sorrell coined the word "frienemy" (part

friend, part enemy)[12] to describe the digital giant that was simultaneously pushing the advertising industry in new and profitable directions
while grabbing a large chunk of his market share. A year later, WPP
had started to make the Web a key plank of its own strategy and had
invested $1 billion of its own clients' money with Google. The tech-
nology giant had gone from being just a threat to being a client and
partner as well, or what Sorrell called a "friendlier frienemy." By 2012,
WPP was spending roughly $2 billion on Google's products and services
and, referencing the popular TV show *Mad Men* about an advertising
company in the 1960s, claimed that it employed "Maths Men as well as
Mad Men (and Women)." He went on to say, "Thus we go head to head
not only with advertising and market research groups such as Omni-
com, IPG, Publicis, Dentsu, Havas, Nielsen, Ipsos and Gfk, but also
new technology companies such as Google, Facebook, Twitter, Apple
and Amazon and then with technology consulting companies such as
Infosys, Wipro, Accenture and Deloitte."[13] In other words, WPP had
embraced digital technology and come to understand the importance
of coopetition. Writing in 2015, he noted: "Take our frienemy Google:
our biggest media trading partner at $4 billion out of $73 billion of bill-
ings in 2015, and at the same time, one of our main rivals too. Xaxis and
AppNexus [belonging to WPP] face off against Google and DoubleClick.
It's a formidable competitor that has grown very big indeed by—some
say—eating everyone else's lunch."[14] That's the kind of multi-layered,
multi-industry co-creation that you and your business are already
involved in, or will be very soon.

HOW TO DESIGN YOUR
SECOND WINNING MOVE

The key message from the four zones is about the inherent dynamics
for capability co-creation across the three phases of transformation. In
the industrial age, companies had well-defined roles, and relationships
among companies were structured around a well-understood logic
of what each party brought to the table. As you map your evolution
along the four zones in the digital era, you'll see that's no longer the

case. Now, coopetition is at the core, pulling and pushing different patterns of relationships between incumbents and digital giants, between incumbents and tech entrepreneurs, and between digital giants and entrepreneurs. It's these interactions that influence the new ways in which businesses are developing their capabilities.

Structuring and managing your portfolio of relationships has always been important. You know that well. You have mostly developed your winning moves within prescribed industry definitions: you have established long-term contracts and followed the clearly demarcated roles of supplier, competitor, and partner. Digitization changes the context for your portfolio of relationships: the *structure* of your relationships changes over time as new digital technologies emerge and mature; the *position* of your relationships changes across the coopetition zones as you make new choices, finance new investments, and set new priorities; and the *relative importance* of your relationships changes as you and the other players make competitive moves. So let's delve deeper into how best to design the second winning move.

Enumerate your capabilities to win in the digital future. By now, you realize that the business capabilities you need to win in a world that is progressively digitizing are different from the ones that you have mastered. To overcome the success traps, you must select the distinctive capabilities that you need to master to win. Consider the three phases one by one. Do you need to become better at scanning and interpreting experiments at the edge of your industry and others? Do you need to develop a more systematic logic to co-invest with partners and develop the business arrangements? Do you need to learn how to better mitigate risk in networks involving multiple partners?

Reflect also on what's needed for you to become a master orchestrator across ecosystems. Will you create and control a software platform? Will you safeguard the data that flows through the networks? Will you design products with digital technology at the core? Will you solve problems for clients in ways that deliver value at a lower cost than ever before? Think widely and write down the capabilities you think will be relevant and valuable in your setting. I have found that the very act of creating this list is liberating for most managers, especially if the setting involves cross-functional experts. You are arriving at your list of the key drivers of revenue and profits as your industry digitizes and your company repositions within and across ecosystems.

Decide what's core (inside your organization) and what's to be co-created (with partners). The second winning move doesn't mean you should design all capabilities with external partners; it behooves you to decide what you can do within your company that gives you a differentiated advantage and what you could co-create with others. The stronger your internal capabilities are, the more likely it is that you will attract stronger partners and the easier it will be for you to absorb the digital capabilities within your organization. Looking at your list of capabilities from step 1, find one best digital partner you could work with in the leader zone (where you are in control) and one best partner in the follower zone (where you let the digital partner define the course of evolution). Chances are that the digital partners in these two zones will be different. Recall that in the leader zone, you are in control and you are working with partners that need you more than you need them. Analyze the pros and cons of locating the capability in one of these two contrasting zones. You may want to look at how your direct industry incumbents have approached the same initiatives and figure out ways to differentiate your own approach, wherever possible. You are essentially developing ways to position yourself differently from your incumbents wherever and whenever possible and also deciding whom to develop longer-term relationships with.

Move selected key capabilities to the co-creation zone. Now, you are looking at how capabilities and relationships evolve and assessing what you could do to move to the upper-right quadrant in Figure 4: this is where you are co-creating capabilities that you couldn't do by yourself. If you are TAG Heuer, could (should) you create a joint venture for connected watches that combines your legendary competence in watch design with Intel and Google's prowess in technology? If you are Fiat Chrysler, could you persuade Google to create a joint entity by pooling your respective resources to create autonomous minivans? Or if you are Levi Strauss, could you see a future offering that takes advantage of Google's Project Jacquard to create seamless clothing with embedded technology? If you are Babolat,[15] the French company that made the first tennis racquet that gathers data and tracks player progress, could you co-create newer capabilities with Under Armour or Ralph Lauren, companies that have made wearable technology to monitor an athlete's health? Like every incumbent, you should examine how pooling your best resources and working with a strong partner with superior digital capabilities could create value in this zone as opposed to the other two zones.

When I ask managers in workshops or education sessions to think of a co-creation partnership that would be high on their list, most eagerly propose joint ventures with one or more of the digital giants such as Google, Apple, Facebook, Amazon, or IBM. When I ask these managers, "Why would they want to partner with you?" and "What do you bring to the relationship that they could not get from anyone else?" that exuberance quickly fades. In other words, co-creation is an articulation of the rationale and logic of mutual dependency and respect for complementary areas of expertise. IBM-Apple is one such joint venture that we referred to earlier with a high degree of mutual dependence and complementarity.

Manage the dynamics of core capabilities. The essential part of this analysis is to think about the dynamics. The importance and impact of the digital initiatives change—partly because your own priorities and preferences change and also because of competitive moves, maturing functionality, and commoditization. Ten years ago, mobile web was in its growth phase; five years ago, social web was in its growth phase. Today, smart home technology and the IoT are growing, whereas the mobile and social webs are showing signs of maturity. Just as reprioritizing digital functionalities is an intrinsic characteristic of digital business transformation, so, too, should you reprioritize your coopet-itive initiatives as you rebalance your requirements for the short and long term. Try to understand what might cause your partners to reposi-tion your co-creative relationship: although you may want to maintain your existing relationship, your partners may want to shift the relation-ship to a different zone as they pursue their own digital agenda.

REMEMBER THAT DIGITAL business transformation lies at the nexus of scale, scope, and *speed*, which means that dynamic evolution is both implicit and also central to your success. You might have been comfort-able with the speed of shifts within your traditional industry, but the speed of digital transformation is faster and ferocious. It is not for the fainthearted. The shape and structure of digital business ecosystems will change even faster in the future. Not only will this affect where you participate in ecosystems and when you orchestrate them, as we saw in Chapter 6, it will also dramatically change the nature of your relationships within these ecosystems. Knowing when to transact, when to lead, when to follow, and when to co-create is the key to your second winning move.

One emerging and particularly dynamic area of change involves the way humans and machines are interacting now and how that relation-ship may evolve in the future. This new frontier may provide you with new ways to orchestrate across ecosystems, as well as new avenues to co-create with others within ecosystems. It's a challenging and contro-versial area, but understanding this particular relationship constitutes the third winning move. This is the focus of the next chapter.

CHAPTER 8

AMPLIFY YOUR HUMAN TALENT WITH POWERFUL MACHINES

WHAT CAN CHESS, an ancient strategy game probably originated in India around the sixth century AD, and *Jeopardy!*, a quiz game conceived by Hollywood producers for television audiences in 1964, tell us about winning moves for digital business strategy? At first blush, these two games seem to have nothing in common; one's a game of skill and tactical maneuvers likened to war and gamesmanship, the other, a test of common knowledge. Nor do they seem to have any relevance for business—except that chess involves thinking through many sequenced moves against a formidable opponent. But both are connected to digital shifts. Both are connected to IBM and powerful machines. Both are connected to how smart humans and powerful machines interact and amplify each other to create new capabilities that couldn't be done otherwise. Both could teach us how to develop the third winning move for the digital business future.

RECOGNIZE THAT COMPUTERS ARE SMART REASONING MACHINES (AND THEY'RE GETTING SMARTER)

Until very recently, the common belief was that a fairly ordinary computer could beat most humans at chess, but it could never beat the very best masters and grand masters. Then, IBM's Deep Blue computer defeated Grand Master Gary Kasparov in 1996 in one game, but the grand master won the tournament. The next year, on May 11, 1997, IBM's computer defeated the grand master in a six-game tournament.[1] That marked an inflection point: computing gave notice to humans in 1997.

IBM wasn't content to leave its experiments in artificial intelligence and cognitive computing there. Most people believed that even if computers were lightning fast at finding answers (think Google or any search engine on steroids), they could not go from starting with an answer to framing the right question for that answer. The assumption was that humans are inherently better and faster at cognitive reasoning through complex paths. Moreover, even if a computer could reason faster than *most* humans, it could not beat the *best* human at *Jeopardy!* That belief was shattered in February 2011, when IBM Watson competed against the two best *Jeopardy!* contestants in the show's history, Ken Jennings and Brad Rutter, and won.[2]

From about 1996 until 2016, only the specialists in the field of machine learning and artificial intelligence knew about the technical developments in computer technology that were behind IBM Deep Blue defeating Grand Master Kasparov and IBM Watson outplaying the two *Jeopardy!* champions. The information wasn't secret, but understanding the developments in computer science required an advanced degree; making sense of what that could mean for businesses required expertise that combined technical knowledge with business acumen. To most managers in most industries, these were experiments at the edge of the computing industry that most executives saw more as "noise" than as "signals" of future things to come.

In 2016, artificial intelligence and cognitive computing (broadly referring to the same general theme for our purpose) have become mainstream, partly because IBM has created a unit to commercialize the IBM Watson technology that puts artificial intelligence, data and analytics, and cloud computing at its core.[3] IBM, as part of reinventing its business for the next phase of its transformation from an industrial-era company that delivered computing products and services to a digital giant at the epicenter of solving crucial problems, signaled its intent to help companies in three major areas: health care, financial processing, and customer service. Computers have gone from being merely efficient tools for automation and the redesign of mundane tasks to complex reasoning machines. This is the reason that it's time for you to take notice. Think about how computing and machine intelligence might allow you to solve complex problems and create new pockets of value for your customers. Think also about how you might be able to augment your industry knowledge with powerful machines and orchestrate certain ecosystems and/or co-create with others within ecosystems. How might reinventing your business with powerful machines at the core transform your organization and the logic of designing your work? In other words:

- How could your strategic decisions be informed and guided by machine intelligence?
- How could thinking of computers as smarter processors of information and knowledge with an ability to reason change how you design work within your business?
- How might putting machines at the center of value creation affect whom you hire to push the frontier of designing next-generation machines?

REDESIGN HOW YOU THINK ABOUT WORK

At the root of cognitive computing, or within IBM Watson's "brain," are hundreds of analytics that process natural language, analyze text, and represent knowledge and reasoning to make sense of huge amounts of complex information in split seconds, rank answers (hypotheses) based on evidence and confidence, and learn from its mistakes. In IBM's parlance, "With Watson, every product and process can understand, reason, and learn."[4]

A three-part question can help you to think about the possibilities.

1. Could IBM Watson[5] replace you in your job?
   You may think:

- "My job cannot be done by a computer."
- "My work involves complex coordination, and that cannot be delegated to a computer."
- "My work is based on years of experience in this industry and a long corporate history, and a computer cannot replicate that."

Like many executives, you may believe that automation is for routine administrative jobs, not high-level strategic thinking, planning, and coordinating jobs like yours. After all, we've been automating jobs like customer support or payroll accounting or human resource processes with computer software and specialized applications for many years now. But there's no magic boundary between jobs that can be automated and others that cannot; cancer cure is not routine, mortgage processing and insurance modeling are not routine, but IBM Watson has focused on all of these. Today's experiments at the edge will become mainstream soon, and that includes computers being able to understand your job and the jobs of all your employees.

2. What parts of your job could IBM Watson do better than you?
   This question is not about IBM Watson taking over your entire job. It's about how powerful computers can support you by becoming your smart assistant.

- What tedious parts of your job could IBM Watson take over?
- How could a smart machine help you streamline or redesign your key tasks and workflows to compete with your digital competitors?
- How could a smart computer help you map the various possible outcomes of several long-term strategies and investment decisions?

You might wonder if the technology has matured enough to process so many different variables, including your own repository of past corporate knowledge. The question is not if, but when. If you do not start thinking about how cognitive computing redesigns key tasks and activities, your competitors will already have embraced this technology and gained its benefits before you begin. We're not talking about a race against machines; we're talking about you taking advantage of cognitive computing to outperform others in your industry. If you continue to have humans carry out jobs that are potentially better done by computers or already carried out by powerful machines inside your competitors' companies, you are already being left behind. Remember, the digital giants have computing at *their* core; why wouldn't you take advantage of powerful machines to guide your own reinvention as a solutions company?

3. How should you redesign your job to take advantage of IBM Watson?
   This is not about just superimposing IBM Watson onto your existing processes. You want to step back to design work across functions, maybe even across different levels of your organizational structure and across companies to take advantage of cognitive computing. For example, if a machine can arrive at a reasonably probable diagnosis better than an average doctor, perhaps this task should be carried out by a machine and a nurse practitioner, freeing up some doctors to spend more time on the parts of their job where they bring the greatest value, such as understanding a patient's emotional well-being or other attributes not quantified or captured on standard medical charts. As a leader, you do not need to know the details of cognitive computing, and you may not know the inner working of Watson's brain, but you want to use it to create an efficient, innovative, and nimble work environment in which machines and humans cooperate and jointly help to create and capture value.

   And you want to anticipate how machines can further help you and your employees learn, process, and decide on your actions. In other words, you want to enable smart humans and "powerful" machines (a term I use to highlight that there are others, including Google's Deep-Mind and other incarnations of conversational commerce machines discussed earlier) to work jointly to create value that neither could do alone. As Microsoft CEO Satya Nadella said: "Computers may win at games, but imagine what's possible when human and machine work together to solve society's greatest challenges like beating disease, ignorance, and poverty... The most critical next step in our pursuit of A.I. [artificial intelligence] is to agree on the ethical and empathetic framework for its design."[6]

We are on the cusp of a major shift in which powerful machines will form the foundation of many industries, and perhaps of society. Previously loosely linked ideas such as machine learning, big data, neural networks, artificial intelligence, and robotics are converging in concrete ways that provide new capabilities. Digital giants such as Alphabet, Amazon, Microsoft, and IBM have seen the power of connecting these formerly disparate strands of computer technology, and tech entrepreneurs such as Uber, Palantir Technologies, Tesla, and Airbnb are putting algorithms and analytics at their core. Your success in this digital future depends on grasping the central role of this system of powerful machines and making them the drivers of your core business strategies and organizational architecture, too.

USE POWERFUL MACHINES TO CREATE NEW VALUE AND NEW CAPABILITIES

When IBM Watson won at *Jeopardy!* in 2011, it had one skill: an ability to understand and respond to questions and answers (Q&A) powered by a focused set of technologies. Today, IBM has about thirty new capabilities in addition to Q&A through application programming interfaces (APIs) delivered via the cloud, and the natural language range has expanded beyond English to Japanese, Spanish, and Arabic. And IBM Watson's ecosystem is growing rapidly. More than five hundred tech entrepreneurs are building applications and solutions, and IBM is acquiring a wide range of companies to support this growth, as a critical part of reinventing its core business. The three-part question that I asked earlier is *not* just specific to IBM Watson and your job, of course. It is simply a lens—admittedly a very good, very visible one—to help you see the impact of this system of powerful technologies on organizations in three ways:

1. What tasks could be *automated*, requiring minimal human intervention?
2. What processes could be *augmented* with smart assistants?
3. What jobs could be *amplified* with active interactions between humans and machines?

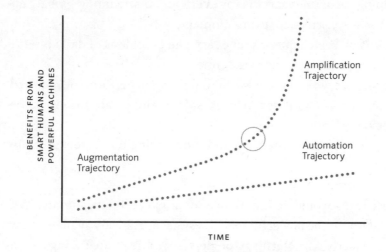

FIGURE 5 Automation, Augmentation, and Amplification

We cannot simply state that all "blue-collar" work could be automated, whereas "white-collar" work could not. We should not. That categorization came from the industrial age that classified work and workers based on production versus administrative tasks, but we know that a lot of so-called white-collar administrative work—production planning, quality control, billing, order processing, customer service, accounting, legal compliance, loan issuance, etc.—has already been automated, thanks to computer technologies. So it's not about looking at the job categories as defined in the industrial era and just applying this system of powerful technologies to them. This winning move calls for rethinking traditional ways of designing work—not just in some pockets or in some functions or in some settings but end to end.

AUTOMATION OF TASKS

In Chapter 1, we talked about your business being at the nexus of scale, scope, and speed, and that makes tasks in the digital age much more complex than in the industrial era. Let's look at some examples:

- Developing recommendations for every video-streaming global subscriber for every screen for every moment.
- Placing ads on search queries for every one of billions of daily interactions across devices, location, and time.
- Monitoring every aircraft engine in flight to then perform the required maintenance without sacrificing safety and while maximizing efficiency.
- Tracking every car on the road and monitoring its performance in real time.

These tasks would be impossible (or at worst, inefficient and ineffective) without the powerful machines that are already available now. Which is exactly why Netflix, Google, GE, and Tesla are taking advantage of these technologies to perform the four tasks above with nearly full automation. The next generation of "cloud robotics," a term coined by former Google employee James Kuffner in 2010, is no longer limited by the size of its memory or its own ability to compute data; it "relies on data or code from a network to support its operation."[7] No matter what your industry is, many of the tasks that you carry out could be fully automated, if not now, then in the very near future.

Studies have suggested that as many as 47 percent of our traditional jobs are at risk, as computers take over automated tasks.[8] But remember that using powerful computers is not simply about reducing the number of employees in your company; it's about competitive efficiency and effectiveness. If you automate faster than your competitors and if you automate a broader number of tasks, you will have a competitive edge. Falling behind is bad enough when shifts in the market are linear; it becomes significantly more problematic when the shifts are exponential. Remember, too, that automation can mean freeing up employees for different work that brings greater value to customers.

AUGMENTATION OF PROCESSES

Although automation is a useful frame, what if powerful technologies could add value to, or augment, some of your tasks instead? Consider the following examples:

- Writing a quarterly management report of your operations by analyzing your areas of success and weakness and identifying ways to improve for the next quarter.
- Filing new patent applications and maximizing the chance of having them granted by reviewing previous applications and indicating precisely how the proposed patent builds on and is different from existing patents.
- Carrying out a first-order diagnostic for cancer.
- Evaluating bids to acquire new television shows by looking at predictions based on past viewing habits of users.

On the face of it, these look like tasks that could not be fully automated, but they can be well augmented by today's technology. In pilot projects, IBM Watson, Wipro's Holmes, and Narrative Science's Quill are already sifting through data and pulling out key numbers, computing key ratios, and plotting them in a comparative format that can serve as the first draft of a quarterly report. If you've recently read a financial news story on the *Forbes* website and did not scroll through to the end, you may have missed this note: "Narrative Science, through its proprietary artificial intelligence platform, transforms data into stories and insights."[9] Simply put, a bot wrote that story. If you allowed powerful technologies to do this for you, what could the impact be on the people you employ? How many people might need to be retrained to work with such machines? What skill mix might you need among your future employees?

In a pilot project, doctors at Houston's MD Anderson Cancer Center have been using IBM Watson to drive a software tool called the Oncology Expert Advisor (OEA).[10] The system can be fed structured and unstructured data—for example, published references on the treatment of breast cancer or lung cancer—in a variety of digitally encoded formats. It then validates the relevance and accuracy of the information and culls anything that is misleading, incorrect, or poorly corroborated. This tool is part live reference manual that is continually updated with relevant information and scientific studies, and part virtual expert advisor for practicing clinicians. From a variety of recommended options, a physician or other attending clinician can choose the one that seems most appropriate for the specific patient. Once the pilot is deemed successful, senior leaders will need to decide whether to continue having doctors work with Watson to diagnose patients or whether to morph the core of the organization towards putting the digital expertise at its core, which has follow-on implications for many aspects of the organization design.[11] Netflix's CEO Reed Hastings has done just that in using powerful machines to query, assess the risks, and understand customers' viewing patterns before bidding on *House of Cards* and creating *Orange Is the New Black*, two highly successful shows. This is augmentation of decisions with algorithms and actions with analytics, and it's a perfect example of why Netflix has an advantage over the traditional TV and media companies and why it is orchestrating many of its own ecosystems.

AMPLIFICATION OF JOBS

Where powerful machines really add value is by working with smart humans to expand the scale and scope of ideas—to amplify them—and this is where you should focus your company's attention. This is where your thinking cycles really help. Amplification depends on two principles: complementarity and singularity. *Complementarity* defines the areas in which machines are superior to humans and then creates governance rules and working conditions to bring out the best combined output and enhanced productivity. *Singularity* anticipates that intelligent computers (computer networks, or robots) will be capable of recursive self-improvement (progressively redesigning themselves) or of autonomously building ever smarter and more powerful machines. In other words, the first principle is about today and the second one is about tomorrow, and they act in tandem.

Why is this important and what does it mean for you and your business? The basic premise is that the design of work is not static. As these technologies acquire more skills and become more powerful, they will shift the nature of work and our relationship with them, which means that you need to create agile workflows and attract nimble employees. More important is that they alter the core talent base of organizations as they become progressively digitized. You may think that Under Armour is a shoe and apparel company, but among its employees are 300-plus software programmers. You might see GE as a power company or an energy company or an aircraft engine company, but its journey with software and analytics has put more than 1,400 programmers on the company's roster. Monsanto may be in the agriculture business, but its strategic focus is more about big data and analytics to maximize yield on the field (beyond the genetic modification of seeds in the bag) and has around five hundred engineers. All the major global automakers have new outposts in Silicon Valley to attract top talent at the nexus of automotive hardware, software, and analytics. There is such demand for engineers, scientists, data scientists, and computer programmers that the US Department of Labor announced in late 2015 that the number of unfilled jobs exceeded 5 million[12] (mainly because of a skills gap.)

So if you want to lead with amplification, where machines and humans accelerate the creation and capture of value, you need to create a climate that attracts the best future talent. What does that look like? It's a work environment in which employees learn at the frontier of how humans and machines work together; it's where software, data analytics, and algorithms are used to streamline decisions, where employees' tasks cannot be done by machines today and they can apply their skills to work with machines to solve some of the profound challenges facing the world in energy, health care, space exploration, transportation and congestion, climate change, and so on.

Is this your company? Is this your invitation to attract the best talent? Is your workplace allowing your employees to carry out their best work? Is the discussion in your company about automation, headcount, and managing for today or is it about amplification and attracting the best talent to solve the world's biggest problems?

AMPLIFY HUMAN TALENT
WITH POWERFUL MACHINES

Artificial general intelligence (AGI) is an emerging field aimed at building flexible, adaptable machines with intelligence comparable to the human mind (and perhaps ultimately well beyond it). It is associated with traits such as consciousness, sentience, sapience, and self-awareness, all of which are observed in living beings. All the digital giants are in the early stages of developing cognitive computing machines. There's IBM's Watson, of course, and its competitor from Wipro in India, cleverly named Holmes. Although Watson (named for IBM's founder) and Holmes may conjure images of Sherlock Holmes and Dr. Watson, there is really no collaboration between the two. In fact, they are in competition with many others, including Microsoft's intelligent personal agent, Cortana; Amazon's Alexa; Facebook's M; Apple's Siri; and Google's Assistant.

So how do we begin to think about this class of powerful machines? I like the distinction offered by the Stanford University scholar on legal informatics Jerry Kaplan: synthetic intellects and forged laborers. According to Kaplan: "Synthetic intellects are not programmed in the conventional sense. You cobble them together from a growing collection of tools and modules, establish a goal, point them to a trove of examples, and set them loose."[13] Watson and friends fit in this category.

The IBM Watson Health ecosystem includes a variety of companies in health and life sciences, such as Medtronic, Memorial Sloan Kettering Cancer Center, CVS Health, MD Anderson Cancer Center, and others, so that the scope of information and knowledge is not confined to silos within individual organizations but is shared to create vibrant co-creative relationships. Synthetic intellects are the foundation of today's trading in financial markets as well as being responsible for directing air traffic, operating the smart grid and mobile networks, and managing security on the nuclear infrastructure. Because they are invisible and built into the infrastructure, we have taken them for granted.

Forged laborers, in contrast, arise "from the marriage of sensors and actuators. They can see, hear, feel, and interact with their surroundings. When they're bundled together, you can recognize these systems as 'robots' but putting them into a single physical package is not essential."[14] For many years, robots were considered fun toys for experimentation. Then companies such as iRobot introduced Roomba as vacuum cleaner, Braava as mopping robot, Mirra as pool cleaning robot, and Looj as gutter cleaning robot. Today's robots are no longer faceless, nameless laborers. New robots—forged laborers—from Rethink Robotics[15] in Boston have names, such as Baxter and Sawyer, and faces. I was impressed with what they could do and how easily they could be programmed. They are different from the previous generation of robots in the sense that they are general-purpose robots and they collaborate with humans; in other words, they work alongside humans and are programmed ("taught") by humans, which makes them powerful allies to automate, augment, and amplify our work.

The first computers were specialized machines, which were custom coded for specific jobs such as running spacecraft for the National Aeronautics and Space Administration (NASA), automotive controls for the major automakers, or word processing programs for home users. Similarly, robots were custom designed for specific applications: automotive production lines, mining pits, farms, warehouses and shipping loading docks, etc. They became more useful when they evolved to general-purpose personal computers, tablets, mobile phones, and wristwatches with a wide range of applications to choose from. In a similar way, we are creating general-purpose robots that can be easily programmed by almost anyone to do many different tasks. And it is only a matter of time before they are networked, too. That adaptability makes robots useful to a wide range of applications.

We are beginning to see collaborative robots work with humans alongside production work or in warehouses. They create new waves of automation that could change the way manufacturers optimize operations and drive efficiency. As they increasingly work together with humans, these sophisticated robots will redefine the nature of work. These collaborations start as experiments at the edge, but they become mainstream very soon as they boost productivity, improve workplace safety, and create a more intelligent working environment with data and analytics at the core. They are challenging business cases about humans versus robots in certain types of work (e.g., pricing airline seats or surge pricing on Uber, recognizing faces in digital images, or deciding on the best fertilizers for different soil conditions in vineyards) and proving that we can have collaborative workplaces in which humans and machines coexist, cooperate, and compete.

CREATE YOUR AMPLIFICATION ADVANTAGE

Digital businesses win because they do things that are at the frontier of powerful machines and smart humans. They are comfortable automating those tasks that should be automated, because not doing so simply makes them inefficient. They also strive to augment those tasks that benefit from using machines as smart personal assistants, as in the case of IBM Watson in cancer care centers, because not doing so makes them ineffective. Most importantly, digital businesses win when they go beyond automation and augmentation to design processes that push the front lines of how organizations could be designed. They attract the best talent and provide them with machines that create multiplicative benefits. If augmentation is additive, amplification is multiplicative.

Think about powerful machines such as machine learning, drones, robots, neural networks, big data, analytics, cognitive systems, and algorithms. They are in their very early stages, just like the Intel computer chips in the first personal computers. They have the power of first-generation Motorola mobile phones, they have bugs and incompatibilities, and they have limited functionality. In short, these technologies are just getting started. But developments in one technology push developments in the others; they are not independent but interdependent. They grow at an exponential rate that the digital giants know how to master. Your challenge, if you're an incumbent, is to look at these developments and transform faster than you have before to keep pace with the change.

I started this chapter by asking what chess and *Jeopardy!* could teach us about digital business strategy. Before 1994, chess grand masters routinely defeated chess-playing computers; by 2004, the results were profoundly reversed. What was the difference? A freestyle chess tournament in which anyone could compete with the help of other humans or computers is insightful. At the beginning of the tournament, human plus machine did better than the strongest chess-specific supercomputer, Hydra. Grand Master Gary Kasparov remarked in 2010 that: "Human strategic guidance combined with the tactical activity of a computer was overwhelming." But the surprise winner of the tournament was a pair of amateur Australian players who used three computers at the same time. According to Kasparov: "Their skill at manipulating and 'coaching' their computers to look very deeply into positions effectively counteracted the superior chess understanding of their grand master opponents and the greater computational power of other participants. Weak human + machine + better process was superior to a strong computer alone and, more remarkably, superior to a strong human + machine + inferior process."[16]

Chess and computers allowed us to begin to understand the complementarity of humans plus machines with supporting organizational processes. But we have a long way to go as we design organizational systems with differing levels of human competencies and unstructured goals. One school of thought is that humans do matter—as designers, coaches, and strategists. Another school of thought is that computers themselves could teach and function as strategists, as Alphabet's experiment pitting its DeepMind computer against world Go board game champion Lee Se-dol recently revealed. According to Demis Hassabis, the co-founder of the AlphaGo computer program designed to play Go, DeepMind relied on "the combination of deep learning, neural network stuff, with reinforcement learning: so learning by trial and error, and incrementally improving and learning from your mistakes and your errors, so that you improve your decisions."[17] It meant that the program could split itself into two halves and, playing millions of matches against itself, learn from each victory and loss. During the course of a twenty-four-hour period, the system could play more than a million times, which is more than the number of games a human could hope to play in a lifetime! That's the impressive power of computing that pushes the frontier of amplification. DeepMind, like IBM Watson, wants to solve the big thorny problem of understanding human intelligence, so the knowledge gained from these experiments with games provides insights that are useful across many industries. In the near future, both schools of thought will coexist.

Over the course of two decades, Gary Kasparov, Ken Jennings, Brad Rutter, and Lee Se-dol—all leaders in their respective games—contested against powerful machines and lost. Their losses, though a blow to their egos, allowed us to see the future potential of technologies in areas far beyond their games. Interestingly, IBM (Deep Blue and Watson) and Alphabet (DeepMind) built these powerful machines with a vision and ambitions that go well beyond winning the games. Mark Zuckerberg at Facebook posted on January 26, 2016, that his team of AI experts was getting close to winning at the ancient game of Go,[18] but it was Alphabet's team that first won. Clearly, the digital giants are racing to develop and redefine the frontiers of *where* and *how* machines could amplify humans. Why is this important?

First, you know that digital giants will both support and challenge your own digital business strategy going forward as you work in and across ecosystems. Second, keen to expand their scope of influence, they have made their programs available to you and other incumbents through software development kits and APIS. We are just getting started with amplification—and the future of work and value creation is not just superimposing technology on old ways of working but fundamentally designing work with computer science at its core.

HOW TO DESIGN YOUR THIRD WINNING MOVE

You, like most industry incumbents, have probably spent an inordinate amount of time thinking about organizational design in terms of structure, processes, roles, skills, and relationships using theories from organizational and social psychology. To succeed in the digital era, you need to start thinking like the digital giants and innovators that embrace computer science as the driving force for organizing. Artificial intelligence may be on your technology agenda, but amplification that catalyzes humans with machines must be on your business agenda. It's time to develop your organizational system at the intersection of humans and machines, and look at the critical skills that need to be mastered. So how do you do this?

Classify your organization into three clusters—automation, augmentation, and amplification. What tasks carried on inside your organization and by your key partners could be automated with currently available and somewhat proven technologies? For example, robots are now becoming widely available and much more affordable: How could Baxter, a $25,000 collaborative robot from Rethink Robotics, change the economics of automating your company's tasks? How could providing such robots to partners in the ecosystem change the logic of your relationships and give you a competitive advantage with automation? Similarly, think about areas of augmentation. Could a joint initiative with a company such as Narrative Science or IBM or Wipro help free up your expert talent from mundane tasks to focus on more value-added areas of innovation and reinvention? Then identify areas at the frontier of amplification—these may not be obvious, since we are in the early stage of experimentation. Do this exercise with a group of senior leaders across different functions, as these three zones may not logically fall within pre-established function-based organizational responsibilities. Think about the functionalities already announced to make use of services such as Watson, Siri, and Alexa through APIS. By looking at other sectors and settings, you may be able to identify areas where and when amplification could become central to your business. If you're in the hospitality industry, you may not have thought much about automation through blockchain technology, but the fact that Airbnb acquired a team of blockchain experts from a tech startup called ChangeCoin in April 2016 should be a wake-up call.[19] You do not necessarily have to acquire that new talent in your own company if you explore creative alliances across industry boundaries.

Benchmark your three clusters against industry incumbents and digital companies. Do a high-level benchmark against your industry peers based on available data. Even crude measures are insightful when you compare the number of people employed in different areas (warehouse, retail stores, customer support, call centers, analytics, and so on). Then look at other strategic moves, in terms of hiring patterns and skill profiles, announcements of joint initiatives or acquisitions and alliances. GE is designing its digital operations by benchmarking not other digital companies but Google, Amazon, and Facebook; it's recruiting talent that would otherwise go and work in one of the digital giants, and its declared ambition is to be a top ten software company, in the company of digital giants. In the automation cluster, you should benchmark against the "best-of-industry" because the comparisons are more direct and relevant. In the augmentation cluster, you should benchmark against the "best-of-breed" to see how you stack up against the best practices for specific jobs and tasks. In the amplification cluster, you are benchmarking more broadly but always trying to push the frontier. Are you staying abreast or falling behind against your industry incumbents and digital giants in terms of combining the power and potential of humans and machines?

Refine and revise the three clusters. Your initial classification of tasks and jobs is only a starting point. As you delve deeper into the frontier of automation, augmentation, and amplification at the intersection of machines and humans, you will find that we are at the early stages. Early adopters of IBM Watson and other such technologies know what's needed to fully absorb such functionality into their organizations. We also know that, over time, incumbent organizations will enhance their ability to absorb such innovations. As new technologies further redefine these three zones, you need to reassess your classification and the talent pool you need. There's no greater challenge for your organization than to understand the profile and composition of talent required to win at the frontier of what MIT Sloan School professors Erik Brynjolfsson and Andrew McAfee call the Second Machine Age.[20] So, in my view, this third winning move puts the digital strategy agenda squarely at the level of senior management and the board of directors.

Reframe your human talent profile. The frontier is, and has long been, humans plus machines. The difference now is that the rate of acceleration of machines has outstripped human talent. So first, you need to bring your human resources department on board to help you design an organization in which machines augment and amplify your human talent. And second, you must realize that you are competing for talent with the digital giants and tech entrepreneurs. Staff from Procter & Gamble, L'Oréal, and Unilever seem to be steadily moving to digital giants like Amazon, Facebook, and Google.[21] It's less about the money and more about creating a business that allows your employees to frame and solve big problems. The digital giants view their human resources as following the power law distribution (also called the long tail): a small percentage (usually less than 10 percent) of people are superstars or hyper performers whose output and value are eight to ten times the average employee; a vast majority, say 60 to 75 percent, are acceptable or good performers with some variations; and the remaining (around 15 percent) are low or unacceptable performers. As Google's head of human talent, Laszlo Bock, writes, "The range of rewards at almost any level can easily vary by 300 to 500%, and even then there is plenty of room for outliers."[22] I believe that Alphabet was created to serve as a talent magnet, its structure designed to attract, nurture, and retain hyper performers within the corporate portfolio.[23] In contrast, traditional companies view employee performance as varying about 60 to 80 percent around the mean, given a normal distribution. So why is this power law distribution important? It means that digital giants are willing to take more risks on people, just as venture capitalists are prepared to back such superstars.

Ultimately, amplifying human talent with powerful smart machines gets at the number and type of people you need in the three clusters of automation, augmentation, and amplification today, recognizing that this relative mix will change over the next decade. Your interactions with the digital giants are more likely to be intense in those areas of augmentation and amplification, and your ability to win in the digital future may be more decided by your legacy human resource practices than by your legacy technology architecture.

USE YOUR THREE WINNING MOVES TOGETHER

I have chosen the three winning moves carefully. Instead of developing a long list of actions that basically cover the essence of good management, I have selected three moves that I believe are not only critical during the period of digital business transformation but have also received much less attention in newer management books. They are related in ways that give coherence to how you use them within and across the three phases.

The first winning move reframes the space of value creation and capture beyond traditional vertical industry boundaries and encourages you to look further and wider at the opportunities for reinventing your business. To credibly solve problems at the intersection of boundaries and address new areas of inefficiency that may lie across industries, you have to be present in many ecosystems. Furthermore, to succeed in these ecosystems, you have to carefully decide where to orchestrate, where to participate, and when to change these roles over time. The result is that you have access to more business models—not only traditional products and services but newer platforms and solutions, too—that allow you to partner with others to gain more detailed insights about your customers and provide personalized offerings at affordable price points.

The second winning move leads from *where* to play (in which ecosystems) and *how* to play (whether to orchestrate or participate) to *who* to play with (the specific partnerships and alliances) to co-create new capabilities and pockets of value, especially in asset-rich industries where digital giants need to work with industry incumbents to obtain domain knowledge and access to channels and physical assets of production and distribution.

The connection between the second and the third winning moves is about using robotics and artificial intelligence to augment and amplify your human talent to create and maintain distinctive capabilities that not only ensure your relevance in the digital world but also attract strong partners for co-creation. So the very logic of organizing for the digital age is how well you are at the frontier of innovation with smart humans and powerful machines. Think of the third winning move as designing the organizational architecture that shapes and supports your business architecture, especially in phase 3, when the traditional industry boundary definitions matter little. These will reflect your positions across ecosystems and how you co-create value within different ecosystems.

These three winning moves, then, are interconnected, both in how you work within and across external ecosystems and in how you structure work inside your organization. In my view, these three winning moves are underappreciated now but will gain in stature. Keep these moves in mind as you begin to think through the logic of adaptation in Chapter 9 and develop your own guiding rules in Chapter 10.

PART 4

YOUR DIGITAL TRANSFORMATION

CHAPTER 9

○———————○

YOUR THEORY OF
DIGITAL ADAPTATION

"**E**VERY ORGANIZATION, WHETHER a business or not, has a theory of the business. Indeed, a valid theory that is clear, consistent, and focused is extraordinarily powerful," wrote Peter Drucker, a noted management guru of the twentieth century, in 1994. He continued, "What underlies the current malaise of so many large and successful organizations worldwide is that their theory of the business no longer works."[1] Drucker might as well have been talking about the digital shifts underway in industries, markets, and companies globally almost two decades later.

WHY YOUR BUSINESS NEEDS
A THEORY OF ADAPTATION

Drucker's belief was that leaders are more focused on the "how to do" than the "what to do" of business. And without a compelling theory of digitization and adaptation, leaders prefer the status quo. You are more likely to follow the rules you know and embrace and adopt digital tools in some obvious areas, preferring to avoid newer, uncharted areas. Without a powerful theory of how digitization transforms companies and industries, leaders fail to see the potentially transformative magnitude, and managers only make incremental changes to their investments. The result is that many businesses are ill prepared for the digital future.

Whereas Drucker was addressing professional managers, Stanford University professor Jim March—one of the leading organizational scholars of the twentieth century—called upon researchers to better understand the *adaptation process* between the exploration of new possibilities and the exploitation of old certainties.[2] This process gets particularly complicated in successful companies, where there's intense pressure to continue with well-known routines of success, because they have worked reasonably well in the past. Even when leaders recognize that the future might not bear much resemblance to the past, they continue along predictable and well-understood paths, not knowing when, where, and how to pivot. Simply put, successful companies trap themselves into their past core competencies precisely when the future may not be an extrapolation of the past.[3] That's the paradox of the success trap that I discussed in the introduction.

The Digital Matrix recognizes that digitization creates a different future from the past, where old rules and moves may have limited use, and thereby calls for a new theory of business. It is intended to help you avoid the four types of traps that I introduced in Chapter 1. The three phases bring the dynamics into focus, and the three sets of players use the three winning moves discussed in Part 3 to create some differentiation. These ideas beg the question: What's the theory of digital adaptation that helps you create your future business in an uncertain and unfamiliar world while you maximize profits from your current one?

WEAK SIGNALS: LEARN TO RECOGNIZE IMPENDING DISRUPTIONS

"The evidence that a major change is needed is hidden among other voluminous signals; it is not loud, clear and unambiguous, like the voice of doom, and it is not welcome information anyway," wrote Professor Igor Ansoff, one of the intellectual leaders of strategy thinking, in 1976.[4]

He was talking about weak signals, those ambiguous and controversial bits of information about the future that are typically hidden amid the "noise" of the prevailing mechanisms you and your organization use to make sense of information, and he was lamenting the fact that these early warning signs are very often overlooked. However, any theory of adaptation rests on an organization's ability to *sense the potential meaning and understand the consequences of weak signals before others do.*

Some weak signals are about fine-tuning today's business, including better understanding trends within the current processes, better targeting marketing messages based on social media listening centers, or better using data and analytics from Twitter feeds, and so on. Increasingly, many companies are getting better at sensing and responding to weak signals in their current operations. But I urge you to focus on weak signals not just to fine-tune today's business model but to better understand how the current business model itself could be disrupted and/or enhanced by digital technologies through the three phases of transformation from Part 2 and the three winning moves employed by the key players that we discussed in Part 3.

Consider some examples. If you were in the packaged software industry in the late 1990s like Microsoft, Oracle, and SAP, software delivered as service over the cloud by the tech entrepreneur Marc Benioff at Salesforce was a weak signal. You might not have fully comprehended the scale and speed of software's migration to the cloud, but you should have tried to understand the meaning of that shift. If you were a big-box retailer like Best Buy, online shopping for appliances via Amazon was a weak signal in the early 2000s. If you were a traditional bookstore chain in a prime retail location in the 1990s, e-commerce was a weak signal. If you have established robust factories and supply chains in Asia today, 3D printing is a weak signal. If you are in financial services, then the blockchain is a weak signal. And if you believe that you have the best human talent in the industry, then robotics and software bots are weak signals for you today.

Ansoff's observations still ring true. Many leaders lose valuable time developing an initial strategic response before they understand the importance of such weak signals, and even more time as they mobilize and reorient their organization towards the new direction. If you are unable to recognize and respond to weak signals in a timely manner, you hamper your company's ability to adapt. Making sense of weak signals is the essence of thinking strategically about your digital future and understanding the possible scale, scope, and speed of these shifts.

Improved automation tools and algorithms are not the answers. You really need to step back and re-examine your assumptions about how your business model will continue through the three phases of transformation. After all, you know about the classic fiascoes such as Xerox,[5] which "fumbled the future" because the parent company failed to understand and appreciate that its far-flung satellite unit, Xerox PARC (Palo Alto Research Center) in Silicon Valley, had invented graphical user interface (Xerox Star) and an optical mouse, the forerunners to the personal computer (PC). Instead, Apple, IBM, and Microsoft recognized these weak signals and developed the first PCs themselves. How, you might ask, could Xerox, a photocopier business facing competitive pressure from other big manufacturers of photocopiers, have seen that these far-out innovations could get them into the computer business? Similarly, how could Kodak have seen the turn of the wheel from photolithography to digital images? Some shifts are just a bit too far removed from the core business for established incumbents to grab onto, perhaps a bridge too far. And just because a company adapts in one time period doesn't mean it has inherent adaptive capabilities and might not fail during a future turn of the technology cycle. Nevertheless, let's look at some companies that are adapting to change right now, as their lessons may provide insights as you embark on your own adaptation journey.

STRONG TRANSITIONS: LEARN TO EMBRACE THE FUTURE

Every company has its own adaptation transition across the three phases. And it is a journey rather than an attempt to reach a specific destination. Those that have succeeded have made key decisions that pivoted them away from their tried-and-tested business logic of the past towards newer pathways for growth. What will set you apart is how well you have understood the importance of digitization and felt the urgency to leave behind your past legacy and embrace the future. Below are examples of companies that have successfully adapted their business, some of which we have already discussed in this book. I have purposely included them again to show that every single major company—no matter what size or what industry—goes through strong transitions at some point. Collectively, this broad sweep of examples should help you develop your own theory of digital adaptation.

Amazon. This company's shift from an e-commerce book retailer to "the earth's most customer-centric company" was part of CEO Jeff Bezos' grand vision and has made it a retailing powerhouse with global scale and scope. In his view, "This is Day 1," and the future is very much ahead for digital business infrastructure. In just two decades, Amazon has outlived most of its original e-commerce competitors, has reached $100 billion in sales, has become more valuable than Walmart (its traditional retailing competitor), and is a leader in cloud computing. It has changed the very definition of scale, scope, and speed in retailing and related areas.

Apple. Steve Jobs' repositioning of the company he'd founded, been forced to leave, and then brought back to rescue from bankruptcy is legendary. The transition involved a shift in scale from producing many versions of a lot of computing products to individual models designed to change the way people live in digital society, first with the iPod, and then with the iPhone and the iPad. This transformation has succeeded by most measures: by mid 2016, Apple was the most profitable and the most valuable company in the world. But it now faces an inflection point as we reach the maturity phase of smartphones.

Comcast. This mass media company could have become just another cable TV and Internet provider (a "dumb pipe"), as several of its traditional competitors are, but by acquiring NBCUniversal from GE, it has increased its scope of content and positioned itself as the gateway for online and entertainment for residential homes and enterprises. Similar adaptation journeys are underway within AT&T (which acquired DirecTV) and Verizon (which acquired AOL, Yahoo, and Hughes Telematics), as they try to avoid becoming commoditized infrastructure providers as the world digitizes even more rapidly over the next decade.

Ericsson. Over the course of 140 years, this communications technology company has evolved from fixed telephony to mobile telephony and has now expanded its scope to deliver network equipment and services to global mobile operators and enterprises. And its competitor, Nokia—despite having been written off as a company that failed to adapt to smartphones from its historical dominance with feature phones—has adapted its scope through the acquisitions of Siemens, Alcatel-Lucent, and Withings to become a leading player in delivering networks for connectivity.

Facebook. The social media company's inclusion on this list may come as a surprise, but in its relatively young life, it has successfully adapted its scope from a social network on personal computers to become the leader in usage and engagement on the mobile web. At the time of its IPO in 2012, it had virtually no presence on the mobile web, but look at the statistics now: more than 1 billion daily active mobile users, of which 850 million use mobile phones only and, monthly, average fourteen hours using Facebook apps. Today, 65 percent of Facebook's video views (competing against YouTube) and 80 percent of its advertising revenue are from mobile devices.

GE. This company's ongoing transformation underscores the fundamental shift in scope from a conglomerate with finance as the connective tissue in a wide-ranging portfolio of unrelated industrial companies to a digital industrial company with software and analytics at the core of new divisions designed to move, power, build, and cure the world. As GE races to build out the Industrial Internet with first-mover advantage, it is looking to attract top talent that might otherwise be drawn to tech entrepreneurs such as Tesla or digital giants such as Alphabet and Amazon. The adaptation aims not only to create the first digital industrial company but also to become a "top 10 software company" by 2020.

Google. This digital giant's adaptation involved a shift in scope from a search-linked advertising business monetized through AdWords and AdSense on the Web and Android on mobile. The new expanded conglomerate model, Alphabet, set up in 2015, allows for many different monetization approaches to be incubated. It allows the company to attract top talent keen on applying emerging principles of computer science to solving thorny and messy problems in many industries beyond search-linked advertising.

IBM. Lou Gerstner was tapped from outside IBM to be the chairman and CEO and rescue the enormous corporate elephant from the brink of bankruptcy in 1994. He masterminded the shift in business scope and organizing logic from highly autonomous business units to highly integrated, customer-focused computing solutions providers. In other words, IBM's clients came to rely on the company's products and services to become more efficient and grow their businesses. Gerstner demonstrated that at least one "elephant could dance"[6] as gracefully as its smaller competitors. Current CEO Ginni Rometty has orchestrated another shift in scope from software and services to capitalize on cognitive computing to further drive efficiency and growth for its clients. Often, size and scale are seen as liabilities for large companies in transformation, but IBM's sequenced transformation shows that these disadvantages can be overcome. This means you can no longer use your large size or long history as excuses to avoid transformation.

John Deere. This machinery company has been complementing its historical mastery in designing and delivering farm equipment by broadening its scope with select acquisitions relevant to the new world of precision and decision agriculture with deeper integration into farms. This expansion of the core competencies, supported by its acquisition of Precision Planting from Monsanto and its deepening relationship with the agriculture giant, could well define the leaders in an industry that is on the cusp of significant digitization over the next decade, as some traditional manufacturers of farm equipment become commodity players, while others move up the value stack with connectivity, information, and insights.

Microsoft. The undisputed king of personal computing is in the midst of an important adaptation to expand its scale for the mobile and cloud world. Its revenue models, honed with licensing fees from its legendary software products—Windows and Office—are changing. By making the Windows 10 upgrade free, moving Office to a cloud-based subscription service, and offering free apps for Google Android and Apple ios, current CEO Satya Nadella has been focused on ensuring that Microsoft remains relevant and central in a digital world that could grow to nearly 50 billion devices connected to the network. Taking the long view and putting aside its missed opportunity at the first turn of the mobile revolution, Microsoft continues its adaptation journey.

Netflix. This company's adaptation involved a shift in scope from the mail-order subscription business with core competencies in efficient logistics and DVD distribution to online video streaming across a wide range of devices and global networks. Netflix had to morph its core to begin influencing the move away from linear broadcast media (television programs shown on specific channels at predetermined times) to on-demand personalized media web (where consumers could access content that is time shifted, device shifted, channel independent, and place shifted within licensing restrictions). It had to develop new capabilities in video streaming from the cloud; refine its data and analytics capabilities; and form new relationships, including with Amazon, which is also its competitor. Netflix's current portfolio of capabilities and relationships is dramatically different from the ones it had as a DVD distribution business, and it is now poised to influence the future of media and entertainment.

Many industries are in the midst of transformation, forcing incumbent businesses to adapt. You are most likely one of them—in the midst of strong transitions. Do you have a compelling adaptation narrative? If you are to be a leader by 2025, you must have your own unique digital transformation story. The essential message of the Digital Matrix is that streamlined adaptation across the three phases is a continuous journey with rapid feedback. As you reinvent your business in phase 3, you'll find yourself engaging in new experiments and colliding with a new set of business models that coexist and morph. Digitization increases the variety of players that invest and interact to shape and evolve existing business models. So adaptation is not a one-stage process with a finish line and a victory lap but a series of multi-stage transitions that occur sporadically but sequentially to ensure profitability in the current period while preparing the foundation for future growth and profitability. Your role as a business leader is to ensure that this adaptation happens without gut-wrenching layoffs, risky acquisitions, and costly recapitalization.

KEY PRINCIPLES FOR EFFECTIVE ADAPTATION

In my view, there's no one model of strategic adaptation that works for every company. My research and experience with companies that have been working hard to adapt with digital technologies lead me to offer the following five guidelines as you craft your theory of digital adaptation.

PRINCIPLE #1: BE PASSIONATELY CURIOUS ABOUT THE FUTURE AND PROFOUNDLY UNCOMFORTABLE WITH THE STATUS QUO

"Someone's going to make your product obsolete," Polaroid founder Edwin Land has often said. "Make sure it is you." The secret is to be curious, or as has been widely attributed to Albert Einstein: "The important thing is not to stop questioning." He believed that it was not special talent but passionate curiosity that allowed him to succeed. These quotes reveal the two drivers of curiosity: a profound discomfort with the status quo and a fascination with the possibilities for the future. Either driver is a valid reason for beginning an adaptation journey.

However, even those who are curious about the future often trap themselves by scanning narrowly and staying within their comfort zone. For example, if your business is agricultural equipment, you are curious about mechanical engineering, but are you also curious about sensors and software? You should be. Yet Harvard professor Clayton Christensen[7] has shown that most incumbent businesses fail to see threats even when the disruptions occur just at the periphery of their served markets. Why? Because they appear to be different from the familiar ways of doing business, and the managers are simply not curious about how such threats could further accelerate and threaten their core.

What we can learn from augmented reality. If you are curious, augmented reality is old news. Its roots go back several decades to military and navigation systems, and it's been in use during Sportvision's football broadcasts since 1998. More recently, wearable technology such as Google's augmented reality glasses, mobile apps that allow students to navigate the books they read in 3D at their desks, and the release of Pokémon Go in July 2016 (which attracted 20 million active users in its first two weeks[8]) have brought this synthesis of the digital world and physical reality into much sharper focus. If you are not already, you should be curious about the possibilities of augmented reality right now. As Sunny Dhillon of Signia Venture Partners notes, true augmented reality "requires computer vision and dynamic mapping of the real world environment" to assess any image or video and understand it as well as we can—all in real time.[9] Although we might not be quite there, you should be asking: What does this ability to integrate the digital into the real and the real into the digital, involving consumers in the co-creation and consumption of products and services, mean for us? What might improvements in machine learning that allow computers to learn and understand without human intervention mean for augmented reality?

Driverless cars might become safer, as they can spot pedestrians with certainty, interpret and avoid road debris, and understand and follow detours. Photo and video archiving might become faster and easier as the computers that run them recognize and sort the images according to ultra-specific details or previously unspecified patterns. Surgical procedures might become more precise, as doctors can draw on expertise from other experts in real time. Fitting room lineups may become a thing of the past, as computers overlay the current year's fashion in the appropriate size and color on your own image. How could this technology affect your industry? What are the weak signals you should track? Curiosity leads to scanning widely across different industries, disciplines, and geographies.

Set up teams with employees who are self-styled futurists and optimistic dreamers—those who see possibilities instead of obstacles. Ask them not to report on trends that can be obtained in standard reports but on likely inflection points: their interpretations, their theories of shifts, and their thoughts about how you should adapt, how fast you need to move, and where the new investments should be.

Make sure this act of questioning the status quo is at the center of your change management program. Design formal mechanisms to understand the assumptions and drivers of success (core competencies) in the past and in the present, but then challenge whether these will lead your company to succeed in the future. And make sure that there are some people within the leadership team with the "hacker

mentality" that Facebook subscribes to—people who are not only profoundly curious about the future but also profoundly uncomfortable
with the status quo. Understand when and how core competencies
could turn into competency traps.

PRINCIPLE #2: DESIGN
EXPERIMENTS TO LEARN

Ralph Waldo Emerson, the prolific American essayist and poet of the
nineteenth century, is widely credited with saying: "All life is an experiment. The more experiments you make the better." And Albert Einstein
is said to have quipped: "No amount of experimentation can ever prove
me right; a single experiment can prove me wrong." In the scientific
community as a whole, it is widely accepted that "observation is a
passive science, experimentation is an active science,"[10] as the French
scientist Claude Bernard opined. But perhaps Jeff Bezos at Amazon best
summarized why experimentation is so important in his 2014 letter to
shareholders: "Failure comes part and parcel with invention. It's not
optional. We understand that and believe in failing early and iterating
until we get it right. When this process works, it means our failures are
relatively small in size (most experiments can start small), and when we
hit on something that is really working for customers, we double-down
on it with hopes to turn it into an even bigger success. However, it's not
always as clean as that. Inventing is messy, and over time, it's certain
that we'll fail at some big bets too."[11] You can learn from experiments
everywhere, especially those with uncertain and complex strands of
convergence with digital technologies. And whether adaptation experiments succeed or fail isn't important; what matters is that they trigger
follow-on experiments to deepen and broaden your learning.

What we can learn from 3D printing. If you are curious, 3D printing is
probably already on your radar because it seems to cut across several
industries. You have seen its use recreationally (to print chocolates or
customize toys) and on a larger scale by manufacturing companies (to
produce proofs of concept and prototypes). For example, Local Motors

in Arizona has demonstrated a full-sized 3D-printed automobile.[12] Boeing has been experimenting with 3D printing aircraft parts across the board for its planes to reduce the cost of maintenance and minimize downtime without compromising safety. The US Air Force has decreed that it will incorporate 3D printing into all facets of its aircraft design and maintenance. Follow-on experiments might include determining if these activities are best carried out alone or in collaboration with other airlines to create "clusters" of maintenance centers, instead of proprietary approaches that may not be as cost effective. Beyond physical products, medical researchers are working on bioprinting human genes on a 3D printer, and Princeton University researchers demonstrated a 3D-printed bionic ear that combines biology and electronics.[13] Although these developments may still be at the edge of current practices, they might usher in dramatic innovations. These are weak signals not only for health care companies but for all businesses, as the learning opportunities from such experiments are limitless.

Similarly, Stratasys has worked with Aurora Flight Sciences to deliver "the largest, fastest, and the most complex 3D printed drone," which can achieve a speed in excess of 150 miles per hour.[14] No matter what your industry, your next questions should be: 1) What new frontiers will Stratasys explore next in 3D printing? and 2) What other types of drones will Aurora design and deliver next? Digital frontiers are exciting and uncertain, and systematic learning reduces that uncertainty.

PRINCIPLE #3:
MASTER THE ADAPTATION CYCLE

Inspired by Charles Darwin's theory of evolution in *On the Origin of Species*, management scholars have developed this derivative quote about organizational adaptation: "It is not the strongest of the species that survives, nor the most intelligent that survives. It is the one that is most adaptable to change." In other words, putting the learning from our experiments into action.

Sense → Learn→ Act → {Repeated Faster}

Successful adaptation requires resources to gain the necessary scale and scope at speed. When you realize that your set of sequenced experiments is showing a clear path forward with ways to mitigate risks, you must seize upon that moment to make the change. It's the big moment to go from allocating minimal strategic resources for a few experiments to committing the organization to a new direction.

What we can learn from the digital giants. For IBM, the moment of action meant creating a separate unit to capitalize on Watson. In 2014, CEO Ginni Rometty committed more than $1 billion for not only research and development but also to bring cloud-delivered cognitive applications and services to the market. The *Jeopardy!* experiment had yielded sufficient insights to be able to create cognitive computing applications for business problems. For Apple, the moment of action was when Steve Jobs extended iTunes beyond music and mobilized the resources to make iOS the center for a family of devices and applications. For Amazon, that moment was when Jeff Bezos allocated investments to make Amazon Web Services a leader in cloud computing. For Google, that moment was when Larry Page committed to the Alphabet structure so that it could mobilize resources as needed to rapidly scale up different initiatives that might need to be nurtured differently than the core search business. At the time, Page wrote: "We've long believed that over time companies tend to get comfortable doing the same thing, just making incremental changes. But in the technology industry, where revolutionary ideas drive the next big growth areas, you need to be a bit uncomfortable to stay relevant."[15] And Alphabet is doing just that with its life sciences unit, named Verily. In addition to producing glucose-checking contact lenses (jointly funded with Novartis), it has also entered into an agreement with Ethicon to form a new subsidiary focused on surgical robots and medical technology for operating room professionals in hospitals.

The end result of scanning and learning experiments is concerted action, even if it means staying the course.

PRINCIPLE #4:
TEST YOUR ADAPTATION CYCLE
AGAINST THE BEST-OF-THE-BREED

As writer Niccolò Machiavelli noted in *The Prince*: "There is nothing more difficult to take in hand, more perilous to conduct, or more uncertain in its success, than to take the lead in the introduction of a new order of things."[16] Which is why it's important to make sure that your adaptation cycle (Sense → Learn → Act) is better than your competitors'. You are not adapting in isolation, and your adaptation moves, especially during the transition from phase 2 to phase 3, set up the new digital business rules that should favor you and increase your chances of winning.

You need to honestly calibrate how well you are adapting along the three phases relative to the other players. First, compare your adaptation cycle to the other industry incumbents by looking at how well you are able to go from sense making to actions: Where are the areas of differentiation relative to your incumbent competitors and where are you weak? I know companies that have an excellent competitive scanning team, but they fail to respond effectively. Second, look further and see how you stack up against the tech entrepreneurs and digital giants. Could you partner with them to accelerate your adaptation cycle? Third, focus specifically on the three winning moves: How good are you with orchestration, co-creation, and amplification relative to the three players? Ultimately, your success will depend on how well you adjust your patterns of these three winning moves to keep up with the best in your field of play.

What we can learn from the us Army. General Gordon Sullivan, who retired as the chief of staff of the us Army, championed fast-cycle learning during the Gulf War. He had small groups gather—wherever they could—to do after-action reviews (AARs) of recent missions and uncover areas of improvements. Participants answered four questions: 1) What did we set out to do? 2) What actually happened? 3) Why did it happen? and 4) What are we going to do next time?,[17] spending 25 percent of their time answering the first two questions, 25 percent on the third, and 50 percent on the "learn to act" final question.[18] The troops that assessed their previous actions were better able to adapt and innovate their tactics.

PRINCIPLE #5:
REDEFINE YOUR PORTFOLIO OF EXPERIMENTS

"Change is the law of life and those who look only to the past or present are certain to miss the future," remarked President John F. Kennedy.[19] It's not a wild stretch to say that he could have been talking about the Digital Matrix. If we assume that change is constant and that you are repeatedly refining your adaptation cycle, then it follows that you will also repeatedly be rebalancing your portfolio of experiments. Just like a mutual fund manager, you must have formal processes to add and subtract experiments to ensure that your scarce resources are wisely allocated.

What we can learn from Bitcoin. If you work in the financial services industry in 2017, you can easily accept that you should already be experimenting with Bitcoin. But when did you first become curious about this innovative payment network and its blockchain technology? When did you first sense its importance and what did you do then? Were you one of the twenty-two major global banks that agreed late in 2015 to jointly create a collaborative "sandbox" lab to design and test alternative bitcoin architectures with distributed ledger technology representing a network of every transaction that has ever occurred?[20] Certainly, this experimentation is a good start, but as the leader of one of these banks, what additional experiments would you carry out alone and with partners to learn faster about the many possible paths in the evolution of digital finance? And if you want to broaden your scope beyond finance into other areas, how could you extend the bitcoin architecture to innovate trusted transactions in settings such as health care or precious metals? This area is ripe for explosion, and your portfolio must be dynamic as well.

The challenge for you as an incumbent leader is to decide where to experiment and with whom (internally versus cooperatively) and when (based on the maturity of the technology). You should also ensure that your portfolio of experiments allows you to learn in ways that help you to better prepare for the digital shifts (which may not be the same reason that your traditional industry competitors approach them). Your ability to win with digital transformation reflects the portfolio of experiments you have already carried out and the ones you have planned for the future.

EXECUTIVES OFTEN ASK me if there's a litmus test for determining how well prepared a company is for digital transformation. The simple and straightforward answer is no. There is no one overarching ingredient, there's no magic bullet technology, and there's no one best way. If you have followed the five principles above, with deep conviction about digitization and its promises and pitfalls, you are likely to be much better adapted than your competition. And that will give you an edge.

CHAPTER 10

○————○

YOUR RULES
MATRIX

"**T**HIS IS DAY 1 for the Internet," wrote Amazon's chief executive officer Jeff Bezos in his letter to shareholders in 1997.[1] Today, I believe this is still Day 1 for digitization, and for all of us who are on the cusp of this transformation. By now, you know that no one company has been held up as a master at digital transformation. Different companies are at different stages with regard to the importance they place on digitization. And some managers see digitization as a tool to improve efficiency without having to change their business model, whereas others see its potential impact on both top-line growth and bottom-line profitability but have not yet been able to mobilize their company as a whole to respond systematically. Your journey, then, is not to follow someone else's well-trodden path but to create your own road map to 2025. So how do you begin?

GO DEEPER: MOVE FROM DIGITAL DABBLING TO DIGITAL STRATEGY

I said at the beginning of the book that your business is already digital—and you have no doubt seen that truth in many different parts of your operations—but your handbook is probably still more or less identical to when your business started. In other words, industrial-age ideas still permeate, and although you may have revised a few things along the way, how you organize your functions, how you think about your industry, how you think about talent, and how you think about success metrics and financial resource allocations have probably remained pretty much the same. You may well have added digitization in small measures in different parts of your organization. Perhaps you have sold your products online and embraced e-commerce, but have you truly become digital? Maybe you have a series of mobile apps, but have you moved from your products being independent to being part of a portfolio of platforms? In all probability, you connect with customers through various social media and you track a lot of interactions that your customers have with other consumers, but have you truly put all this data to use to become more solutions focused? And maybe you have benchmarked yourself against the familiar companies in your industry. Are you or they finding new business models and creating new value at the intersection of your industry and others? Are you or they working with new partners in competitive and cooperative relationships within a network of companies where previous definitions of labels and roles blur?

You may be looking at the developments that the digital giants have demonstrated. Facebook's Aquila Internet drone, which runs on solar power, made its test flight in June 2016 to explore how best to connect the 7 billion people on earth to the digital frontier. If you are in the telecom industry, you'll have noted the immediate implications for your business. But even if not, you have probably considered how this technology could accelerate high-speed data connectivity to parts of the world that have not been touched so far by digitization. How might that affect your business model and your ability to reach new customers? You may also have been looking at virtual reality in different forms—from Google's simple and inexpensive cardboard box to Microsoft and Facebook's more expensive headsets. If you are in the training business, you may already be experimenting with this technology. But even if not, what could it mean for the digital-physical interaction of experiences of different kinds?

Similarly, you may have looked at a variety of innovative developments that entrepreneurs have pitched to you to get early- or late-stage investments. Maybe you've considered machine learning or robotics or ways to integrate the private and public cloud to make your operations more efficient and secure. In other words, you know many of these individual developments, but what do they mean for your business? How should you approach your digital transformation? That's the crux now and the focus of this chapter.

If you are just beginning your digital transformation journey, familiarize yourself with recent writings. Columbia Business School's faculty director of executive programs, David Rogers, lays out five domains—customers, competition, data, innovation, and value (CC-DIV, pronounced see-see-div)[2]—that are to be integrated to create value. Another useful primer is the digital transformation compass introduced by MIT researchers George Westerman and his colleagues. Four phases—framing the digital challenge, focusing investment, mobilizing the organization, and sustaining the digital transition[3]—set the context for taking action.

Now, apply the rules that emerge from the Digital Matrix. When I say, apply the rule(s), I do not mean that there is a rigid set of steps to follow. The whole purpose of strategic thinking, as the digital giants and tech entrepreneurs have shown, is to think beyond obvious boundaries to outmaneuver your competition. Yet the individual rules are worth explaining: they highlight the key principles and analysis that you must undertake. Each rule is a building block, and you weave together the ones that make sense for your strategic thinking now; then, when conditions change or when you're confronted with newer opportunities or more serious challenges, you revisit them to select a different set of rules to create a new strategy.

DESIGN THE RULEBOOK THAT'S RIGHT FOR YOUR BUSINESS RIGHT NOW

In Chapter 2, I introduced the Digital Matrix with the idea that digital transformation can be seen through the actions of three players

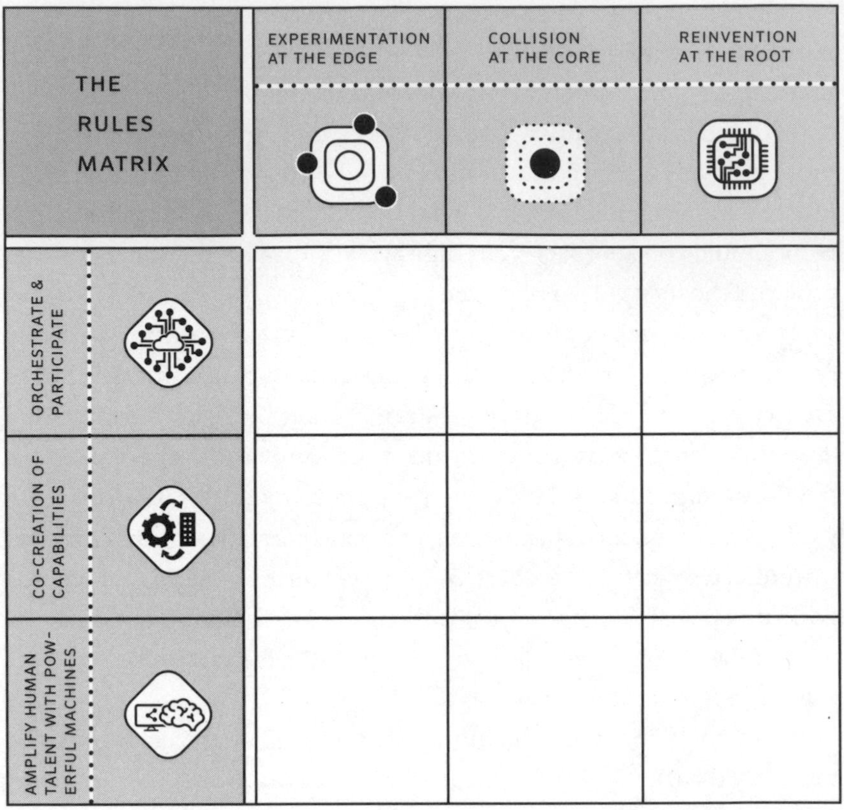

THE RULES MATRIX		EXPERIMENTATION AT THE EDGE	COLLISION AT THE CORE	REINVENTION AT THE ROOT
ORCHESTRATE & PARTICIPATE				
CO-CREATION OF CAPABILITIES				
AMPLIFY HUMAN TALENT WITH POW-ERFUL MACHINES				

FIGURE 6 The Rules Matrix

across three phases. Figure 1 (page 46) illustrates this first core representation. Now I want to introduce a second matrix, the Rules Matrix. Instead of looking at all three players, this Rules Matrix maps the three phases of digital transformation against the three winning moves *for industry incumbents only* and gives rise to nine rules (Figure 6). This exercise is equally valuable if you are a manager in a tech entrepreneurship or a digital giant, but my focus here is specifically on incumbents.

Many management books end with a set of rules, prescriptions, and guidelines. In many cases, these are either very general or specific lists of dos and don'ts that often seem disconnected from the main themes of the book. The Rules Matrix is different in that the nine rules follow directly from the discussions in the previous chapters. Again, these are not a rigid prescription: they raise questions and guide action items rather than providing pat answers. That's because the insights the rules offer are the result of your company's internal and external conditions, which change. That is, the insights you get today will be different than at a later time. As well, often executives on the same management teams gravitate to different rules, which reflect where they sit in the organization and what their particular viewpoint is. This is not a bad thing, and it won't paralyze your ability to take action; making the time to examine these different viewpoints will help you customize your rulebook.

You may be familiar with social media command centers such as the ones that the American Red Cross, Cisco, Dell, Pepsi, and Salesforce have designed in recent years. Your company may even have its own bank of screens streaming real-time data that show when your brand is mentioned in different social media forums (Facebook, Twitter, Pinterest, Instagram, etc.). I like to picture the Digital Matrix command center as a similar setup but designed to track and analyze the responses to your key moves through the three phases, integrate this information with feedback from other sources, and coordinate your follow-on actions and reactions. Imagine this command center as that place where you also track in real time the key moves that the digital giants, tech entrepreneurs, and your traditional competitors are making to transform their strategies and possibly position themselves ahead of you. The command center provides you with the information you need to write your own rulebook, establishes your priorities and positions along each of the nine rules, and serves as your guide for action. I'll describe the nine rules and then show you how to put these rules into practice. For each rule, I provide examples and in some cases link back to previous chapters to reinforce the point. But first I want to give you some ideas about how you might approach using the Rules Matrix.

CHOOSE YOUR ENTRY POINT

Unlike many management tools, it does not matter where you start with the Rules Matrix. Ultimately, you must get to know each rule well so that you can apply it to your context. But to begin, read through all nine rules and think about how they link together in your setting. Don't make any decisions, just note which one or two rules jump out at you. That's your first entry point.

A single rule

If one or two rules in particular seem to speak to you, it's because you have been thinking precisely about the issues and challenges represented by those specific cells. Perhaps you have been invited to a meeting with a digital giant to discuss possible collaboration and you may be wondering how best to think about your role in such an alliance. (Focus on rule #4.) Or maybe you are thinking of reaching out to a digital giant such as Facebook, Alphabet, or Amazon to explore joint initiatives, but you are unsure what specific expertise you bring to that potential relationship. (Look at rule #5.) Or possibly you have been asked to look at acquiring an emerging digital company that has successfully patented a set of innovations that could potentially challenge your historical business model. (Highlight rule #2.) I have been in meetings and workshops with managers who described just these situations; these were new situations to them, and they are likely to be new situations to you.

You can bring your best management thinking, approach these situations analytically, and try to make the best decision. But these are not one-off situations and, consequently, nor are the decisions. You have to develop rules to deal with them because they will occur again and repeat more often as digital giants and tech entrepreneurs link with you in deeper and increasingly complex ways. In short, focusing on one single rule that addresses an important and immediate agenda item in your company right now is a great starting entry point. Work towards mastering that rule fully with regard to your company's needs right now.

A set of linked rules

Remember, though, that the rules are not isolated from each other, as the Rules Matrix makes clear. All the rules in a particular *column* are linked by the phase of transformation in which they occur. All the rules in a particular *row* are linked by the winning move they represent. So instead of beginning with a single rule, you might choose to start with, say, a set of rules that deal with experimentation (#1, #4, and #7). Or, instead, a group of rules that are relevant to, say, successfully navigating the moves associated with a collision at the core (#4, #5, and #6).

If your competitors have raced ahead to reinvent themselves for phase 3, you may want to look at a group of rules. For example, GE has been actively positioning itself to lead a digital-focused Industry 4.0. If you are Bosch, Siemens, Boeing, ABB, and Mitsubishi, you might prioritize rules #1, #4, and #7 to make sure that you are appropriately positioned in the ecosystems that will usher you and others into this world of platforms and solutions. Concentrating on a group of related rules that address an important lag in your company right now is a decisive starting point. Work towards completely mastering all the rules in that phase or that winning move to accelerate your transformation and ensure your continued competitiveness.

All of the rules, in order

If no one rule emerges as particularly important, you may be someone who likes to see the whole holistic picture before delving deeper. In this case, locate where you are along the three phases. Pick the rule in that phase that speaks to you most clearly and start there. If, for example, you are in an industry that has not yet entirely grasped the full power of digitization, but select companies are thinking further ahead and stepping up, start with rule #1 and work through the rest sequentially. If you are John Deere or Monsanto or another player in the agricultural industry, you might take this approach.

Similarly, you may already be part of the way down the road to digitization but want to take a systematic approach to looking at what you have done and how you can be more strategic going forward. If you are an incumbent, such as GM, Ford, or Honda in the automotive sector, which is clearly in phase 2, you might start somewhere like rule #4 to assess how to strengthen your alliances. Then go forward through the other rules—and remember to cycle back to the first three. The reason to go back to the first three is simple: the actions and interactions are dynamic. Just because you face collision today does not mean that the window of experimentation is closed for this industry. In this situation, familiarizing yourself with all of the rules and systematically reviewing and applying your strategy for each one may be the perfect entry point.

Wherever you begin, with one rule or all of them, remember that the rules vary in importance to you over time. Neither your company, nor the industry you are dealing with, nor your collaborators and competitors remain static, so the order in which you follow the rules may not be sequential. And it will certainly change over time. The nine rules are the same for all businesses, but where you begin and in what order you weave together the rules as they become relevant to you is where your application of the Rules Matrix differs from others. That's your unique rulebook; you can customize and adapt it on the fly as circumstances change. The better you understand the nine rules and can apply them successfully, the more likely you are to win in the digital future.

UNDERSTAND THE NINE RULES

As you review the rules, think first about each one's overall intention. That's the key if the Rules Matrix and the idea of the winning moves evolving through the various phases is new to you. If you have some experience with these ideas, or as you put them into action, think not only about your company but about how your direct competitors could be taking advantage of these rules together, too. Could they be in a better position to link with digital giants and tech entrepreneurs? If so, how should (and could) you neutralize that advantage? At that point, your competitive tactic may not be to respond directly at the level of each rule but to look at all nine and strengthen the ones that play to your advantage.

Rule 1. Evaluate your roles in experimentation ecosystems

One of the key changes during your digital transformation journey is the change from a self-focused firm to one that invites and involves complementors in ecosystems. To apply this rule, select one or more ecosystems that can help you learn how to craft this kind of strategy. Note that the ecosystem(s) you select in this experimentation phase may not be the ones you seek to participate in or orchestrate but the ones that might raise reference points for discussion. At this early stage, you are observing, and there is no pressure to orchestrate unless it is an area of high value to you.

For instance, GE initially focused on Facebook. In its goal of becoming a "platform and applications company" for industrial machines and systems, or Facebook for Industry, GE first observed consumer Internet platforms, social network platforms, and app stores to analyze how their ecosystems evolved. Similarly, IBM is focused on the blockchain. In its vision of the Internet of Things, the blockchain is "the framework facilitating transaction processing and coordination among interacting devices."[4] As "a shared, trusted, public ledger that everyone can inspect, but which no single user controls . . . it can be amended only according to strict rules and by general agreement."[5] As a result, it can be used to enhance security, protect privacy, clarify identity, and serve as a building block for smart contracts between smart devices across companies and across industries. Look to see who's stepping up to orchestrate the different ecosystems and understand what the effective orchestrators bring in terms of distinctive strengths that allow them to earn this right.

Note that not participating in any of the emerging digital ecosystems shows that you are resigned to following the business rules set up by others; that decision should be taken seriously after due consideration, not by default.

Actions for rule 1.

- Observe the horizon for important digital trends and invest early in areas that have not yet shown proof of impact but appear promising. (Has an innovation shown promise? Yes. You should probably invest.)

- Participate in experiments in a broad cross-section of general-purpose ecosystems and seek to orchestrate in a few selected domains of potential high value to you. (Is this area of high value to you? If yes, seek to orchestrate. If no, participate.)
- Emphasize experiments that challenge and complement your business models. (Is this an area that directly challenges or complements your business? If yes, prioritize it.)
- Complement what you may be doing inside your own company, including acquisitions. (Is this an area that helps co-create capabilities? If yes, prioritize it.)

Rule 2. Explore capability co-creation options

Another fundamental shift in your digital transformation journey is breaking free of the orthodox thinking that you should not work with your competitors. The digital world is about complex dynamic interconnections between companies. To apply this rule, select one or two ecosystems in areas other than your industry and observe the interactions between traditional competitors. Note that the ecosystem you select in this experimentation phase may not be of high value to your core business but might raise reference points for discussion. At this early stage, you are observing, and there is no pressure to co-create unless it is an area of high value to you. Then test which partnership arrangements will best help you transform your business by strengthening your existing capabilities and building new ones that will help you solve problems for customers and reinvent yourself for the digital world. Wherever possible, aim for capability building in the leader and co-creation zones—and don't shy away from coopetitive relationships: learning to manage them adeptly will give you a competitive edge.

Actions for rule 2.

- Scan widely beyond your immediate industry to understand emerging patterns of capability co-creation when three types of players work together. (Would TAG Heuer's approach to partnership—working with a digital giant and/or a tech entrepreneur to advance its digital capabilities—have applicability for your business? If yes, explore your options.)

- Examine possible futures and select the ones that seem the most likely for your current industry and the ones that will allow your company to differentiate itself and add value. (Would being the first incumbent in your industry to have a digital version of your product or service give you an advantage? Would partnering with one or more of the other players allow you to advance your digital knowledge and capabilities? If yes, seek to co-create.)
- Enter into experiments that will take you from passive observer to active participant. (Is there another player—even a traditional competitor—who would benefit from your knowledge and you from theirs? If yes, invest in co-creating a product or service that adds value for both of you. Be creative and innovative in your partnerships: look outside your industry or shift a current relationship.)

Rule 3. Examine the intersection of human talent and machine power

Venture capitalist Peter Thiel says, "As we find new ways to use computers, they won't just get better at the kinds of things that people already do; they'll help us to do what was previously unimaginable."[6] To apply this rule, start by tracking smart technologies that are redefining efficiency and innovation. Actively explore the next frontier of work and, by extension, of organizing itself. Look to other industries, to the digital giants, to experiments with collaborative bots, and examine how your organization could structure its work and use powerful machines to bring out the best of your company's human talent. To learn and adapt, invest in selected experiments that have high value for your company.

For example, Goldman Sachs, the global investment banking firm, has been experimenting with an internal software platform, called Marquee, to create risk management and analytical tools. An app called SIMON helps clients learn about structured investments and execute their own transactions. As more users sign on, Marquee uses the additional data to fine-tune its accuracy and customize its clients' offerings.[7] The end result could well be that "Goldman will still have the chief product of a bank—money to lend and invest—but... the ways in which customers get access to that money will rely more on software and less on the bankers who traditionally delivered Goldman's services."[8] Note that this rule is not just about deciding whether to substitute robots for people; it's also about understanding what the shift means for productivity, innovation, and organization. In this case, Marquee changes service delivery and makes Goldman a financial solutions company.

Actions for rule 3.

- Assess how to take advantage of automation and powerful machines proactively, not only to offload the mundane, repetitive parts of work to bots and thinking machines, but to specifically attract the best talent to push the boundaries of what we think of as possible. Goldman has 9,000 tech-savvy employees out of 35,000 (Would creating your own digital platform—internally or with partners—bring added value to your customers and your company? Would having more digital-savvy employees help your business extend your scale, scope, and speed? If yes, it's time to invest.)

- Look for areas of work in which machines could amplify human talent in ways for your firm to differentiate itself and add value. (Would your customers benefit from being able to access information about your product or service from anywhere, at any time, customize it, and get answers to questions about it in real time? Would your company benefit from more information about how your customers use your product or service, and from more immediate feedback? If yes, invest now.)

- Involve your human resources department in describing the type of talent your company will need to attract and how they will work with powerful machines to deliver value. (Would your organizational structure and workflows benefit from being part of "the sharing economy,"[9] in which teams and data are widely connected internally and with expertise in the cloud, flexible, and able to pivot quickly? If yes, start here.)

ALL THREE OF these rules focus on scanning and interpreting the experiments underway at the edge of your traditional industry and others, and carrying out your own tests to position yourself for the future. You are deciding on your roles in experimentation ecosystems, you are delineating possible and probable areas of co-creation with partners among the three sets of players, and you are describing the future of your company's work at the intersection of human talent and machine power. With a broad exploration agenda supported by appropriately allocated investment to learn fast from mistakes (yours and others) and follow-on actions, you can move from observation to investment and thereby set yourself up well for the next phase of digitization. The next three rules are about that second phase—collision at the core.

Rule 4. Engage ecosystems as transformational triggers

Most companies fail to recognize when they need to make change. I believe that to successfully transform from an industrial-age company to a digital one, you must begin the shift within your business (using your own internal capabilities), but the trigger to accelerate and move forward powerfully is participating in (and orchestrating) ecosystems (assembling new capabilities with external partners). This is especially true as traditional technologies collide with digital technologies at many levels—products, processes, services, and organizational architecture—where no one company can amass the required capabilities on its own. Ecosystems allow you to understand the nature of the transformation required, decide on the core capabilities that you must develop, and rely on partners to support you in other areas.

For example, Netflix is orchestrating a multi-layered video-streaming ecosystem that includes partners producing a set of Netflix-ready devices, such as televisions labeled "Netflix Recommended TV" that can be remotely upgraded; third-party devices such as Apple TV, Roku, Chromecast, and others; cable TV set-top boxes from global cable TV and satellite TV operators; game consoles, such as Sony PS4, Xbox One, and others; and Blu-ray players; personal computers, smartphones, and tablets. Financially, Netflix could not have produced all this hardware;

instead, it tapped into a vibrant community of hardware manufacturers. Moreover, the ecosystem that Netflix put together included content agreements from leading global entertainment studios and wide-ranging Internet service providers (ISPs) to ensure optimum viewing across different bandwidth speeds. More critically, as we saw in Chapter 7, it worked with (and still continues to rely on) Amazon for cloud capability. In other words, Netflix relied on an ecosystem to achieve scale and scope at speed.

Actions for rule 4.

- Learn to understand the network structures[10] of ecosystems. Understand who are emerging as the leaders at the center of these hubs and examine how the ecosystem structure is evolving. (Would joining emerging networks of incumbents and digital giants help you understand and test new business models? If yes, then join. Do you see opportunities to create new ecosystems to help you develop guiding principles about coexisting and morphing as you transform your business? If so, focus your thinking here.)
- Set up a separate unit inside your corporation to innovate and incubate new ideas. (Would giving employees dedicated time, money, and authority to pursue digital ideas help your company advance its digital transformation and engage in a more focused way in ecosystems? If yes, this coexistence approach might be worthwhile.)
- Examine your role in ecosystems—as an orchestrator and a participant—to accelerate your digital transformation. (Would working with others on projects involving new digital technologies help to accelerate your technological knowledge and capabilities? Would orchestrating ecosystems around your own experiments allow you to add scale, scope, or speed? If yes, start to morph your focus in this direction.)

Rule 5. Navigate your preferential partnerships

You, like most companies, probably have a set of principles that guide your relationships with supply chain partners, contract manufacturers, marketing companies, and so on. But those focused on the future, say with venture capitalists or joint patent applicants, are probably more ad hoc. Structure and manage them like a financial portfolio. Navigate your roles of orchestrator and participant by selecting relationships that give you the requisite capabilities. Rebalance and reprioritize the different relationships as conditions change.

Take the case of Corning Inc. The industrial glass and ceramic producer is aiming to transform its glass surfaces into sophisticated electronic devices by embedding them with sensors. To do this, it has deepened its relationship with Samsung, which buys Corning glass for its phones and tablets. In addition, it has entered into equity investments with View, Inc., a Silicon Valley pioneer that is innovating smart ways to regulate heat and glare into rooms from external glass surfaces. Both are preferential relationships: Samsung to coexist with today's business model and View, Inc. to possibly morph its core for the future Glass Age. Corning's ability to morph its business and to become truly relevant to higher-value participants in digital business will depend on these two relationships and others that it must enter into to win.

Actions for rule 5.

- Select a set of preferential relationships that help guide your shift from the traditional to the digital world. (Would working closely with one or two targeted partners allow you to maximize your current business model or accelerate your digital transformation and acquire specific knowledge that will help you differentiate or create new value? If yes, approach those partners now and enter into the relationships to maximize their role.)
- Rebalance your portfolio of relationships periodically. (Have some relationships outlived their value? Do other developing ones need more of your time? If yes, proactively adjust your preferential partnerships to achieve scale and scope at speed and morph your business.)
- Maintain your current co-creative relationships while cultivating new relationships in the ecosystems that will lead you towards your digital goals. (Would forming new partnerships in different industries help you co-create more broadly, better sequence your moves, or become

more adept at navigating partnerships? If yes, coexist with your current partners and morph your business with new ones.)

Rule 6. Augment smart humans with powerful machines

If there's one area of digital business transformation that deserves serious discussion and debate in most companies now and in the near future, it is this rule. Historically, discussions about technology have occurred in different parts of your organization, but to be successful step back and look at the entire set of tools that make up the frontier of smart machines and develop a road map to adopt them and adapt your ways of working. Think, too, about the types of talent you need to hire going forward.

For example, PayPal uses a "detective-like methodology" of deep learning combined with human expertise to fight fraud and turn a profit.[11] Such functionality has now been incorporated into highly successful tech startups such as Palantir Technologies, a software and services company specializing in big data analysis for such clients as the Central Intelligence Agency (CIA) and the Federal Bureau of Investigation (FBI), as well as oil and gas companies, health and life sciences companies, and financial services companies. Digital giants have raced ahead with productivity gains through sustained applications of machine learning and artificial intelligence, and conversational bots and cognitive computing are becoming mainstream. The key to success with this rule is not simply overlaying technologies on the traditional organization but fundamentally restructuring workflows and work teams to integrate them, allow more sharing, and use computers and humans where each is best suited.

Actions for rule 6.
- Establish areas of differential expertise by juxtaposing powerful technology and human skills, not simply by buying the best technology from vendors. (Would having your workforce and all of your digital systems focused on solving a big thorny problem allow you to accelerate your transformation to lead in the digital world? Would it allow

you to create value for your customers and new opportunities for your employees? If yes, then invest in those areas to morph your business.)

- Develop and continually refine areas in which humans + machines could add differential value to your business and your work processes. (Would a highly integrated organizational and technological workflow allow you orchestrate ecosystems, co-create innovations, and bring new value to customers? Would it allow you to clearly articulate your future moves? If yes, focus your efforts here.)

- Roll out your new vision in areas where the impacts are likely to be significant and meaningful. (Would combining smart humans and machine learning allow you to analyze and interpret data more powerfully; use it to innovate more subtly or more deeply; and refine your products, services, and solutions more quickly and more personally? If yes, act now.)

RULES 4 THROUGH 6 focus on managing relationships during the phase in which business models are colliding and the key challenges for incumbents are to coexist (hybrid traditional and digital) as long as possible while preparing to morph their business infrastructure, platforms, and ecosystems for phase 3. The next three rules come into play during that third phase.

Rule 7. Design your new digital ecosystems

In digital ecosystems, products and services are linked to platforms and solutions pushed by different manufacturers and pulled by different consumers. During this period of profound uncertainty when you are reinventing your business, know that you have charted a course with business principles that you believe best position you to win, so clearly articulate where you orchestrate the relevant ecosystems (including what capabilities help you to do so) and where and when you participate in complementary ecosystems—and then do it. But keep scanning for new opportunities (and threats) and continue adapting to new technologies, new customer wants, and your own innovative ideas for the future.

As part of its transformation into a digital industrial company, GE has designed an ecosystem around its Predix software platform, which monitors more than 50 million data elements from industrial machines to minimize downtime and maximize efficiency. For this purpose, it is a founding member of the Industrial Internet consortium and has pulled together such diverse partners as Accenture, Vodafone, Pivotal, AT&T, and Cisco. By orchestrating this ecosystem, GE has positioned itself as a leader in industrial data at scale; can grow this ecosystem exponentially and bring more value to its partners by attracting traditional incumbent companies in industries such as lighting, water, aviation, mining, oil and gas, health care, and transportation to join this ecosystem; and continue to develop the capabilities necessary to succeed through 2025 and beyond by leveraging its expanding portfolio of partners.

Actions for rule 7.

- Scan widely and work with a variety of players to identify problems that could be solved with digital technologies for the different sectors of the economy and for society. (Would orchestrating ecosystems with all three types of players in your industry and outside it help you solve some big thorny problems for your traditional customers and for others? Would doing this allow you to capture new pockets of value; generate more revenue; add scale, scope, and speed to your offering; and transform your entire business? If yes, dive in. This is your goal.)
- Link your products and services to other products and services on different types of platforms at many layers of your business as part of solutions ecosystems. (Would being more highly linked to the other players at more levels of your business help you to transform your business and take a larger part of the value pie? If yes, why wait?)
- Constantly fine-tune your offering and ensure that your partners' and customers' needs are being met. (Would solidifying your position as a respected orchestrator help you to keep innovating and adding value

for your customers and for society? Would framing your actions this way help you to stay nimble, keep the best talent, and position you for relevance in the digital future? If yes, commit here.)

Rule 8. Co-create new business capabilities with preferred partners

In all of the digital business models I have analyzed, all of the leaders create new capabilities with others—often those considered "competitors." They know that in a fast-changing world, no one has the required resources to develop the capabilities acting alone and that the risks are better mitigated through joint co-creation. Remember, your preferred partners are not static: keep an eye out for new opportunities and divest yourself of partners that no longer bring value to co-creation.

In June 2016, BMW announced a partnership with Intel and Mobileye (a tech entrepreneur from Israel that had initially aligned with Tesla but then allied with this initiative) to create a fully autonomous self-driving car platform named BMW iNEXT.[12] This three-way partnership is in the best spirit of the three types of players underlying the Digital Matrix: it combines BMW's knowledge of the car industry, Intel's expertise in computer systems and electronic design, and Mobileye's experience with building camera sensors for cars. The idea is to build the platform, share it with other car manufacturers, and generate revenue and new capabilities through licensing agreements and other deals.

Actions for rule 8.
- Explore widely how other industries are reinventing for the digital era with powerful ecosystems. (Do any of their orchestrators provide pointers to help you frame or solve problems? If so, get started on structuring and orchestrating similar ecosystems in your industry.)
- Understand the preferred relationships that could serve as the foundation for your reinvention. (Do you have a critical set of preferred relationships that bring complementary capabilities to define your role as a key problem solver and solution deliverer? If yes, why wait?)

- Recognize that preferred relationships evolve to reflect change in market dynamics and the shifts in capabilities required to navigate them. (Are there different partners with new capabilities to complement what you bring to the ecosystem? Are there new opportunities for co-creation? If so, rebalance your portfolio of relationships and refine the preferred list to include those that will help you frame and solve problems.)

Rule 9. Differentiate with computer capital and human capital working together

Digitization is more than ads, apps, and automation. It's about putting computer science at the core of designing your business. So the final rule is about finding the best talent to work with machines, not in preconceived categories but to create innovations that will solve the core problems of industries, economies, and societies. Your organization, and all organizations, will become talent magnets not just to execute but also to explore, not to just implement but to innovate and inspire. This is the frontier where machines + humans working together are continually coexisting and co-creating, and it changes where you invest and who you attract, what you patent and how you develop competencies that outperform competitors.

Let me cycle back to why you should keep digital giants on your radar. Every one of them is pushing the digitization frontier with smart humans and powerful machines. Amazon with cloud, drone, conversational bots (Alexa), and robotics in warehouses; Microsoft with emotional computing and conversational bots; Apple with conversational bots (Siri) and possibly cars and transportation; Alphabet with DeepMind and artificial intelligence driving search and advertising, as well as health care and transportation; IBM with Watson and artificial intelligence finding new solutions in health care and other industries; and Facebook with artificial intelligence in social interactions and conversational commerce. Digital giants differentiate themselves from other IT companies in how they take advantage of powerful machines: they will define employment as we reinvent the relative contribution of human capital versus computer capital.[13] As Peter Thiel remarked, "There's room in between for sane people to build a vastly better world in the decades ahead."[14] How you approach this rule may well define the core of your organization as you reinvent it for the next phase.

Actions for rule 9.

- Proactively understand how computer capital and human capital are interacting now in different settings to frame and solve problems. (Do any of those practices provide relevant and valuable pointers for your reinvention? If so, take advantage of amplification now so that you don't have to play catch-up in your ecosystems later.)
- Examine how amplification could allow you to orchestrate ecosystems in which you have previously been a participant. (Could you restructure the ecosystem to orchestrate using machine learning, algorithms, and analytics? If so, develop the synthetic intellect as your edge and seize the advantage to lead with it to solve some big thorny problems.)

RULES 7 THROUGH 9 focus on streamlining your relationships and using them and technology to reinvent your business and organizational structures by framing and, ultimately, solving problems in society. Your goal is to differentiate your company from others, offer new value, and position yourself to compete in a fully digital world.

DEVELOP YOUR COMPELLING NARRATIVE

Competitive conditions vary by industry and geography, so not all of the nine rules will be equally applicable for all companies, but it is crucial to look at all nine and understand their roles and place. No matter what your entry point, sooner or later, you will be using them, or you will be competing against someone who is using those rules, because there's only one end goal for every company: to develop a well-reasoned and analyzed transformation road map by applying the relevant rules to reinvent your business for the digital world.

The rules help you to develop specific actions: in phase 1, how to experiment to observe and determine what precise level of investment to make in those few promising areas of relevance to you. In phase 2, how best to coexist with traditional business and digital business operations and, more importantly, at what specific scale, scope, and speed to morph your core so that you don't get left behind during the critical window of transformation. In phase 3, how your industry may be positioned in ecosystems that emerge at the intersection of industries, and how your relevance could be challenged by other ways of framing and solving the big thorny problems. For example, what's the role of insurance in a world of self-driving and autonomous modes of transportation? What's the role of biopharmaceutical companies in a world of wellness? What's the role of telecommunication carriers in a world of drone-provided connectivity? The problem domains change, the set of plausible solutions change with digital technologies applied to solve them, and the traditional competencies simply get marginalized.

The rules also help you develop a compelling narrative, a picture of the kinds of big thorny problems that digital technologies could solve and the implications for your business going forward. It is this persuasive message about the opportunities and challenges that lie ahead—and most importantly, the ways in which you are seeking to differentiate your offering and create value for customers—that you want to deliver to your organization. And you also want to determine the levels of investment required to maintain that differentiation against the digital giants, the tech entrepreneurs, and your own industry incumbents. So sharpen your message by highlighting where and how you add differentiated value relative to others.

ACT NOW!

Every company faces its own unique digital inflection point. Past business models are showing signs of ineffectiveness. New directions and options with digital technologies are daunting. Savvy digital upstarts are intimidating. However, the choices made at such points in time matter because they define the subsequent avenues and follow-on options. Just as Kodak, Nokia, BlackBerry, and Sony either failed to see or simply could not adapt when faced with these inflection points,

I believe that many more companies will miss their windows of digital transformation.

Remember that the Digital Matrix sees digitization as an ongoing evolution that influences how important problems are framed and solved in economies, industries, and society; how companies are created and re-created; and how organizations are designed and redesigned to deliver superior value to consumers. Just as the Industrial Revolution was neither a single point in time nor a single technology in a single industry or geography, digitization will have far-reaching global impacts in the twenty-first century and beyond. The bottom line is that it's not too late to step up to the digital transformation space, but the rate of change will be rapid and the risk of inaction will soon be more expensive than the risk of experimenting and getting started. Act now!

Clarify your long-term objectives. Look at the initiatives that your organization has undertaken. How comprehensive are they? How could they be refined? The Digital Matrix and the Rules Matrix help you to match your long-term goals with your current capabilities to know which next moves to take. For example, we've seen that the Accor hotel group in France has instituted digital hospitality programs that target clients, employees, and partners with systems and data. What follow-on actions would you suggest? Would amplifying human talent with powerful machines allow the company to co-create new capabilities and provide customers with new pockets of value? Or might orchestrating one or more of the hospitality ecosystems allow the company to increase its internal efficiency and customer service, thereby expanding the scope of the program? Both are valid directions, but one will probably be a more logical extension of the company's position at the intersection between the three phases, the three sets of players, and the three winning moves.

Similarly, we've seen that Siemens has announced its next47 unit to focus on disruptive ideas to accelerate new digital technologies. What specific actions might make the most sense for Siemens' long-term strategic innovation? Should it seek to co-create in the leader zone with specific tech entrepreneurs or digital giants in order to differentiate itself from other incumbents? Might it be better for the company to develop its digital systems so that it can take the company in new directions? Again, both are possible directions, but one is likely to be a better fit for Siemens in the current environment. The point is that you can be in any company, in any industry, nearly anywhere in the world, and you will find ways to apply the nine rules and iterate fast to gain advantage. Your success lies in doing this more adeptly than your competitors and your collaborators.

Think about new winning moves. We should not be blind to the possibility that another winning move might emerge—perhaps a move focused on an area such as security, privacy, or identity. These three areas are becoming increasingly important and deserve extensive discussions of their own. A winning move must truly challenge the direction of your digital transformation and offer you new insights for inventing your business model at the scale, scope, and speed you need to win. The three phases are enduring and evolving; the existing three winning moves are important business rules that could be supported by additional attention to specific technical areas. The current command center has nine screens, but future command centers might have twelve or more.

Let me say again that there's no one universal theory of digital business and there is no one best way to make the transformation. However, the digitization of your business is the most important item on your company's strategic agenda. You may have started along the transformation journey in different parts of your organization and now need to coordinate these piecemeal initiatives.

MOBILIZE YOUR COMPANY
FOR DIGITAL TRANSFORMATION

History is filled with cases of businesses that have failed during pivotal points in time. Past successes and timely transitions do not automatically guarantee that you will choose the right window to adapt and succeed in the future, and knowing the principles of effective leadership does not ensure you will put them into practice. It takes the right combination of people to invest in experiments and interpret their results, to determine when to pivot and change course, to engage others in making the necessary changes. Smart humans + the Rules Matrix is your key to digital success.

But, ultimately, the transformation of your company is a collective action, one carried out by like-minded professionals with a common purpose. You must start with your enthusiastic energy, and then invite and mobilize others to join you on this journey.

ASSEMBLE YOUR DIGITAL LEADERSHIP CORPS

The formula for success is simple: strong leadership plus a diverse, skilled, and curious group to make change happen.

WANTED: MANAGEMENT LEADERS WITH SCIENTIFIC MINDS TO GUIDE ADAPTATION

How your company's leaders embrace and work with the five adaptation principles (Chapter 9) gives you a robust clue about how well your company will fare in implementing your new rulebook. First, you want managers who are curious and desire change (or as the late Intel CEO Andy Grove titled his book about exploiting crisis points: *Only the Paranoid Survive*[15]). Second, you want leaders who clearly accept that experimentation is a systematic way to discover deeper insights—even at the expense of fast failures. Third, you want leaders who connect experiments to actions and commit real resources to them. Fourth, you want leaders who benchmark their adaptation ability against others in the industry, and then more broadly against digital giants and tech entrepreneurs. Finally, you want managers who have put in place mechanisms to continually rebalance the portfolio of experiments so that their strategic direction is informed and influenced by these experiments. The proof of your company's ability to adapt is in the actions. Are you willing to lead by example?

To be an effective leader, you must understand the need for a *theory* of value creation and capture, where the theory is dynamic and fast evolving with digitization. You must have a scientific mind that is curious, eager, and prepared to explore new dimensions of possibilities and examine new models and alternative approaches. This does not mean you have to be a management scientist with an advanced degree in management. Far from it: you need to understand emerging digital business trends, connect the dots, and weave the different threads into a coherent way forward. You need to translate that theory into resource reallocations for digital business transformation. And you need to work with others in the organization to constantly assess your priorities, tweak your rulebook, and work towards implementing it. In fact, every industry, every company, urgently needs this kind of leader. Is that you? Could that be you?

WANTED URGENTLY: A DIVERSE GROUP WITH COMPLEMENTARY VIEWPOINTS

As we've seen, it's not enough to just enhance your skills or competencies about digital business. You have to act. And acting means assembling a corps of leaders who are as excited about the digital future as you are, and equally frustrated about the lack of substantive and coordinated ways to deal with digital transformation. They are your fellow warriors. Your digital dream team is composed of dreamers, designers, doubters, and doers.

{Digital Leadership Corps = Dreamers + Designers + Doubters + Doers}

Collectively, they represent different viewpoints but share the same vision. They need not be separate individuals, but make sure that your team includes these four complementary characteristics. The dreamer's crazy ideas need the designer's discipline to translate them into different experiments. The doubter makes sure that the ideas are worthy of investment and commercially viable. The doer builds the organization and ecosystem necessary to implement the ideas at scale. In assembling the team, look for the following characteristics:

1. *Those who command respect for their domain knowledge and expertise but respect other points of view and are curious about other domains.* These individuals bring intellectual diversity to the group.
2. *Those who have deep conviction and strong viewpoints but are persuaded by data and analysis.* These individuals give credibility to the recommendations arrived at by the team.
3. *Those who have demonstrated a track record of success with strategic projects.* These individuals give stature to the initiative.
4. *Those who have the self-confidence to respect the past but are prepared to challenge orthodoxy and the status quo.* These individuals bring fresh insights.

Collectively, this team believes that the future is not a linear extrapolation of the past. These leaders recognize the power of digitization and have the shared confidence to deal with the inevitable ambiguity. They are not impressed by frameworks but persuaded by well-reasoned insights. They're drawn to debating how to frame the questions as much as they are to delving into the assumptions that support specific answers. Most importantly, they're responsible for the full cycle from imagination to implementation. They're your company's *digital leadership corps*. And they work with the Rules Matrix to get started on the digital transformation journey.

USE THE RULES MATRIX IN THREE STEPS

There are three simple steps to putting these rules to use in your company. Once you've assembled your digital leadership corps, have them work through the Rules Matrix systematically. It's useful to have each person do this exercise individually and then compare notes as a group. Use Figure 7 as a template for each rule and compute the three values—Importance, Proficiency and Gap. This will help you keep track of your answers and determine your priorities.

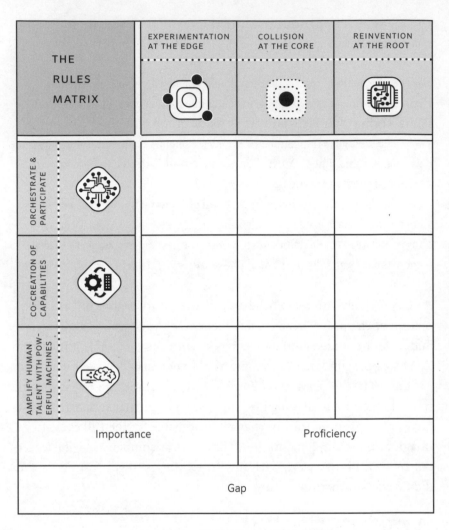

THE RULES MATRIX	EXPERIMENTATION AT THE EDGE	COLLISION AT THE CORE	REINVENTION AT THE ROOT
ORCHESTRATE & PARTICIPATE			
CO-CREATION OF CAPABILITIES			
AMPLIFY HUMAN TALENT WITH POWERFUL MACHINES			
Importance		Proficiency	
Gap			

FIGURE 7 Your Digital Business Rulebook

STEP 1: ASSESS THE IMPORTANCE
OF THE RULES TO YOUR BUSINESS

For each of the rules, assign a score of anywhere between 0 (not important) and 10 points (a top priority). For example, you might assign a score of 4 for rule #3 if you are observing and experimenting with automation in this early phase but still a ways off from implementation. And you may give rule #6 a score of 7 because you can foresee that amplification is or will be an important strategy to shift your business model in phase 2. And you may give rule #9 a score of 10 if you believe that augmentation will be highly important as your company strives to develop a strong foundation with digital business in phase 3. The scores are your assessments based on your prioritization. Work through the nine rules sequentially, filling out the grid as you consider each move. Once you have assigned each rule a value, review the scores and adjust them as you see fit to reflect where the company is presently located in terms of importance across the nine rules in Figure 7.

STEP 2: EVALUATE YOUR PROFICIENCY WITH THE RULES TODAY

Assess how proficient your company is in mastering each rule and executing it, now and as you look ahead. For each of the rules, assign a score of anywhere between 0 (not proficient) and 10 points (highly proficient). This is not an absolute scale; measure your company's proficiency relative to others that are jockeying against you. Use the following guideline:

0 to 4: Weaker than industry incumbents
5 to 8: Parity with others (including all three sets of players)
9 to 10: Leader of the pack (across the three sets of players)

Again, work through the rules sequentially, filling out the grid as you consider each move. Review the scores and adjust them as you see fit to reflect where the company is presently located.

STEP 3: CALCULATE YOUR IMPORTANCE-PROFICIENCY GAP

Now comes the important part: subtract the score for proficiency from the score on importance for each rule.

[Importance – Proficiency = Gap score]

You may get negative values, and these are perfectly acceptable. Once you have calculated these gap scores, arrange the rules in order from highest gap score at the top to lowest gap score at the bottom. This is your list of priorities for digital business transformation. The ones at the top deserve your attention immediately.

Look at the nine cells of the Rules Matrix and see where you need to pay more attention and when. You are in a much better position to honestly and fairly assess your level of proficiency with these rules—e.g., Can you truly pull off orchestrating ecosystems of importance during phase 2 or redesign the company at the frontier of machines and humans as you go towards phase 3?

The gap that you have computed is your assessment as an executive, who is very vested in seeing the organization succeed. But you're not alone; there are others like you in your organization. Now let's involve them.

TEST YOUR RESULTS

Now that you've arrived at your set of priorities, discuss them with your colleagues in the digital leadership corps. You are but one person—albeit an important change agent—in your organization, and yours is but one perspective. Your colleagues may have different assumptions and views on the promise and pitfalls of digitization. Test how your priorities stack up against theirs, either formally or informally.

- How do your top 3 to 5 rules compare with the group's as a whole? Where's the biggest variance in the group? Try to understand what underlying assumptions and points of view are contributing to the variance.
- What are the top 3 to 5 areas of distinctively superior proficiency? Is there broad consensus? If there are wide variations in your assessments, try to understand what underlying assumptions and points of view are contributing to them.
- What are the top 3 gaps? Pay attention to the rules where the gaps really stand out and develop actions to close them as the first step in applying the rulebook.

Even if you are in a position of authority to undertake changes, having the input of others is valuable for two reasons. First, it is useful and insightful to understand how others think about the importance of the nine rules and the degree of the gaps. Second, you cannot transform your company on your own; you need a cadre of change agents. If you are not in a position to influence changes by yourself, this allows you to mobilize support from others; such support may be needed to unleash transformative changes from the trenches. As a team, you are now aligned towards the future. Your job is to pull your organization away from its legacy thinking and towards the digital future. The success of your company depends on you.

The digital leadership corps is your force for transformation. They are not an ad hoc committee working off the sides of their desk; they are your future leadership team committed to designing the foundations for future success. Unleash these leaders and your new rulebook—your strategic vision based on your collective assumptions, experience, aspirations, and analysis—now to guide you confidently towards the digital future. The future is exciting for those who see the possibilities; it's dark and dismal for the rest. I hope you see the potential and promise. There's no time to waste. There's no reason to wait.

NOW, YOU MUST LEAD

As Gandhi said: "You must be the change you wish to see in the world."

"Do you see a future with the company where you work now?" If the answer is no, then hurry to find a different company where your talents and aspirations may be better aligned! If the answer is yes, act *now*. The biggest challenge for organizations is when the most passionate and energetic people stay quiet.

You have two options. One is to wait for senior managers to decree that digitization is important and announce a portfolio of digital transformation initiatives, and then, figure out where you could play a useful part and contribute. The other is to lead proactively. Initiate change from the middle of the organization out and be the change agent, the digital transformation champion.

You do not need to create a powerful "burning platform" memo, such as the one that Stephen Elop wrote at Nokia in 2011[16] (such memos are severe, often a bit too late, and invite rebuke). You simply need to persuade like-minded people in your company to join forces with you. You need to convince the skeptics with data, analysis, and prototypes as appropriate. You need to build bridges across functions, divisions, and geographies, as the case may be, to find possible centers of competencies.

You may wonder: Where do I start? My suggestion is for you to start with the Digital Matrix and think through the nine screens on the control panel, and then move onto your own assessment of the nine rules. You have rated their importance and your assessment of the organizational competences. You are convinced of the risks if your organization does not embark on the top two or three rules immediately. You have also rallied a group of your fellow coworkers to do their own assessments. Now, you have some colleagues whose top rules match yours. Despite minor variations in the scores, I am sure that the overall narrative about your company's digital transformation road map is quite consistent.

Here are some ways for you to take the next steps.

1. Find a champion at the highest level of the management team who shares your vision and viewpoints and could be your backer and sponsor.
2. Conduct an internal survey using the nine rules to check the pulse of the company with as broad a group of employees as you can. What are the emerging patterns from the trenches?
3. Use the results to write a "call to action" memo. Highlight the costs of inaction—staying with the status quo. Highlight the costs of slow actions—responding slowly and with halfhearted energy. Specify the dangers and risks, draw connections back to the weak signals, and develop implications and constructive guidelines.

4. Identify areas where you could begin to change and find other colleagues who will join forces with you to embark on a transformation program.
5. Start small, experiment, iterate, and demonstrate the way forward.

Change agents do not wait for directives; they seize opportunities, because they understand their importance and have a keen sense of urgency. They also know that these change programs are not minor rearrangements of roles and responsibilities, but that they have fundamental implications for top-line growth and bottom-line profits. You are not embarking on change for the sake of change. You are doing it to lead the company on a successful trajectory of transformation. You see yourself as part of the elite pack of the future leaders.

I have written this book with two objectives. One is to educate and guide a generation of managers to think about the digital future in a systematic way that goes beyond glorifying a set of technologies to be an important driver of big shifts. This reflects my calling as a scholar and an educator. The other objective is to prod a subset of readers to actively play constructive roles in digital business transformation of companies. There's no more important challenge facing companies today. Those that do not see the transformative power will find themselves disrupted and morphed in ways that we have not seen so far. So I hope you step up to be the transformative leader. This reflects my calling as a consultant.

Kevin Delaney, the editor in chief of the digital native news outlet *Quartz*, wrote: "I don't think anyone should have the word 'strategy' in their job title."[17] I tend to agree with that reasoning: there is a wide distance between strategy and execution, and assigning some employees with strategy in their title disempowers others. I could make the same argument for the word "digital." No one should have that word in their job title. It creates an exalted status for some and disempowers others from thinking about ways to leverage the power of digital technology. I am confident that by 2025, there will be no difference between digital and non-digital when it comes to functions, processes, business models, and industries.

You, as a reader, irrespective of your job title, independent of function or business line or hierarchical level, do have a chance to play an important role in the transformation of your company. You can play this role if you are passionate about taking advantage of the power of emerging technologies and see your larger purpose as guiding the company through this inflection point. I hope you do and that this book plays a small part in your success.

ACKNOWLEDGMENTS

○————————○

It's cliché to begin by saying, "Writing a book is lonely, but no book is ever written without help from others." But that cliché is true in my case. Although I have been the sole author of academic papers and articles for management journals, this is the first book I've been the sole author of, aimed at professional managers. The ideas I present here have been influenced by my research over the past three decades, and many of these research projects have been carried out jointly with academic colleagues and former doctoral students. That collaboration has undoubtedly helped me arrive at and refine many of these ideas, as have the countless opportunities I've had to translate and interpret the results of my research for practicing managers and would-be managers (my MBA students).

My entire professional life has been characterized by the interplay between academic research and teaching in the broadly evolving arena of digital business strategy. At the very beginning of my career, I was fortunate to be involved in the Management in the 1990s research project at the Massachusetts Institute of Technology (MIT), which was led by Professor Michael Scott Morton. His mentorship, along with that of the late Jack Rockart, helped me get started, and I am grateful for that. I then had the opportunity to apply these emerging findings and frameworks as an academic expert in consulting projects carried out by the MAC (Management Analysis Center) Group, which subsequently became Gemini Consulting, the Index Group, and later, CSC/Index. And over the past fifteen years, I have had many occasions to present and test my ideas with a range of companies in different parts of the world: at the top management forums in London and Johannesburg; at in-company workshops for ABN AMRO, BP, BT, Ericsson, Merck, Boston Children's Hospital, Kaiser Permanente, Homeplus (Tesco) in Korea, FedEx in Brussels, L'Oréal in Paris, MAS in Sri Lanka, and Honeywell in New Jersey; at IBM events at Oxford University and Boston University; at IBM workshops for client teams in Brazil, Dubai, and Paris; at United Nations–sponsored workshops in Riyadh, Brunei, and Doha; at Harvard Business School, MIT Sloan School, London Business School, INSEAD, and the Nelson Mandela African Institute for Science and Technology in Tanzania; and at alumni events at my own alma maters: the Indian Institute of Management, Calcutta, and Boston University. After nearly every one of these engagements, someone invariably asked if I had a forthcoming book on the subject. My response was always that I had a book project in mind but no clear completion date yet. Now, that book—the one that summarizes my ideas in words, not just in PowerPoint and Keynote slides—is here.

I would be remiss if I failed to mention some of the special professional collaborations that have helped my thinking over the years. My association and friendship with John C. Henderson in the last three decades has been invaluable and important. From our initial work together on the strategic alignment model at MIT in the late 1980s through today at Boston University, he has always both encouraged me and challenged me to push my ideas beyond the obvious. Working with him has helped me to develop many of the ideas discussed in this book, especially alliances and partnerships in ecosystems, including capability co-creation. I have always enjoyed my interactions and collaboration with Nalin Kulatilaka at Boston University's Questrom School of Business, and he has helped me over the years to refine my ideas while presenting at various events and co-teaching sessions with him at the EMBA program at Questrom. His ability to probe deeper and see broader connections to finance—especially risk, options, and value as important frames for strategic resource allocation—has helped my work. My three-way collaboration with John and Nalin has truly made my academic career at Boston University special and rewarding. Michael Lawson, who recently retired as the senior associate dean at the Questrom School of Business, was instrumental in creating the research environment that made this collaboration and learning happen. Finally, my associations with Bala Iyer (previously my faculty colleague at Boston University and now at Babson College) and Chi-Hyon Lee, my doctoral student at Boston University (now at George Mason University) stimulated my ideas on competing in business networks and ecosystems in Chapters 6 and 7.

Beyond academia, several other professional collaborations deserve special mention. Jo Guegan, who spearheaded the IT strategy practice at Gemini Consulting in the early 1990s, saw the power and value of the five-level business transformation logic and the strategic alignment model. My ideas have benefited from working with him over the years, including most recently as a member of his Global Advisory Board at Canal+ in Paris. My ideas on experimentation at the edge evolved from some of the initial discussions during those meetings. My partnership with Rick Chavez, which started in the early 1990s, has continued until today. I was part of his advisory board at Viant, an Internet strategy firm, during the dot-com boom, and that experience helped me see the transformative power of the Web in shaping business strategy. I continue to learn with him as he develops his strong points of view on demand-centric pull as a driver of digital business. Clearly, the ideas of reinvention in Chapter 6 have been influenced during discussions in many different forums. Jim Ciriello, now at Merck IT (previously at Lucent Technologies), saw the complementary ideas that John, Nalin, and I bring to this arena of digitization of industries, markets, and firms early in 2001, and he funded the Boston University Institute for Leading in a Dynamic Economy (BUILDE) to help us focus on the power of mobile technologies. My interest in the role of different webs as driving new business models and organizing logic emerged from that initiative, and he continues to be a strong sounding board to test ideas as he applies some of them in his work at Merck. I think the health care industry is on the cusp of profound digital transformation—as I have mentioned in the book—and I plan to extend the Digital Matrix with a particular focus on health care in my future work. Steve Newman, who recently retired from Ericsson as the director of executive development, has seen my ideas evolve from 2000 until today, as I have engaged yearly cohorts of high-performing executives to think about the power of digital technologies and their importance to Ericsson and its clients. He also read an early draft of this book and provided useful pointers. Chris Newell, director of learning at Boston Children's Hospital, has seen my ideas evolve from his days at the Lotus Institute in the 1990s through to his current role, and he has helped me sharpen these ideas to make them relevant and applicable to a wider audience.

The book is the outgrowth of the master class on IT strategy that I have taught at the Questrom School of Business, where for the last four years, we examined how digital giants impact and influence the strategies and actions of many different industries. I have learned a lot from my students. The ideas have been slowly evolving over the last few iterations of the course, and I hope my former students—now practicing managers—get to see the full set of ideas integrated and actionable in the Digital Matrix.

Writing for professional managers is daunting, especially in today's world of simple rules and catchy phrases. It's particularly challenging for someone who has mostly written for an academic audience. For my maiden venture, I decided to forge a nontraditional path by working with LifeTree Media rather than a more traditional university-based press. This is because I believed in publisher Maggie Langrick's vision to help first-time authors craft compelling books for a broad audience. And I am so pleased that I chose this avenue. Maggie saw the power of my ideas and the timeliness of the topic, and encouraged me to put my ideas on paper (to be precise, digital ink!), and she connected me with Lucy Kenward, who took my garbled prose and terse treatise and made it far more readable and accessible than I could ever have done alone.

And Shirarose Wilensky polished the manuscript. Setareh Ashrafologhalai illustrated the figures superbly to complement the text, and Paris Spence-Lang ably managed the project and kept it on schedule. Thanks to all the members of the LifeTree team.

Several friends and colleagues graciously took time off from their summer to read early drafts of the book. I am grateful to Mike Lawson, Steve Newman, Bala Iyer, Mohan Subramaniam, Ben Bensaou, Michael Scott Morton, John C. Henderson, Nalin Kulatilaka, Omar El Sawy, Bhaskar Chakravorti, Rick Chavez, Jim Ciriello, Mel Horwitch, Brinley Platts, Georges-Edouard Dias, Heraldo Sales-Cavalcante, Don Bulmer, Alejandro Martinez, Chris Newell, Jo Guegan, and others. Although I have not been able to incorporate all their suggestions here, I have reflected on them seriously, and some may find their way into my other writings.

Finally, I gratefully acknowledge the financial support provided to me by the David J. McGrath Professorship that I have held at Boston University since 1998. Thanks also to Ken Freeman, dean at the Questrom School, for seeing the digital future, selecting it as a strategic area of study for the school, and creating an environment in which to develop new ideas at the intersection of research and teaching.

NOTES

PREFACE

1. Nicholas Negroponte, *Being Digital* (New York: Alfred A. Knopf Inc., 1995).
2. See, for example, Michael A. Cusumano and Richard W. Selby, *Microsoft Secrets: How the World's Most Powerful Software Company Creates Technology, Shapes Markets, and Manages People* (New York: Touchstone Press, 1998); Eric Schmidt and Jonathan Rosenberg, *How Google Works* (New York and Boston: Grand Central Publishing, 2014); David Kirkpatrick, *The Facebook Effect: The Inside Story of the Company That Is Connecting the World* (New York: Simon & Schuster, 2010); Brad Stone, *The Everything Store: Jeff Bezos and the Age of Amazon* (Boston: Little Brown & Co., 2014); Eric M. Jackson, *The PayPal Wars: Battles with eBay, the Media, the Mafia, and the Rest of Planet Earth* (Washington, DC: WND Books, 2012).

INTRODUCTION

1. Mark J. Perry, "Fortune 500 Firms in 1955 v. 2015: Only 12% Remain Thanks to the Creative Destruction that fuels Economic Growth," *American Enterprise Institute* (blog), accessed July 19, 2016, www.aei.org/publication/fortune-500-firms-in-1955-vs-2015-only-12-remain-thanks-to-the-creative-destruction-that-fuels-economic-growth/. The companies on both lists include IBM, General Motors, General Electric, Procter & Gamble, and Boeing. American Motors, Zenith Electronics, and National Sugar Refining were on the list in 1955 but not in 2015, whereas Facebook, Apple, Amazon, Netflix, Google, Microsoft are on the 2015 *Fortune* 500 list but did not exist in 1955.
2. Richard Foster, *Creative Destruction Whips through Corporate America*, executive briefing for Innosight, Winter 2012, accessed July 18, 2016, www.innosight.com/innovation-resources/strategy-innovation/upload/creative-destruction-whips-through-corporate-america_final2015.pdf.
3. Interbrand, *Best Global Brands* 2015, accessed July 18, 2016, interbrand.com/best-brands/best-global-brands/2015/ranking/. By way of comparison, in 2000

Interbrand ranked Apple at #36 and Amazon at #48; Google and Facebook did not make the top seventy-five. By 2008, Apple had improved to #24; Amazon had dropped to #58 and Google was at #10 (it had entered the top 100 list only three years earlier at #38).

4. Here I am deliberately referencing but updating Marc Andreessen, "Why Software Is Eating the World," Wall Street Journal, August 20, 2011, www.wsj.com/articles/SB10001424053111903480904576512250915629460.

5. My initial interest in looking at the impact of information technology on business operations began in the 1990s, as a participant in a collaborative research initiative at MIT Sloan School of Management. The results of the MIT study are published in Michael S. Scott Morton, ed., The Corporation of the 1990s: Information Technology and Organizational Transformation (New York: Oxford University Press, 1991). My research findings were subsequently presented as N. Venkatraman, "IT-Enabled Business Transformation: From Automation to Business Scope Redefinition," MIT Sloan Management Review, Winter 1994, accessed July 18, 2016, sloanreview.mit.edu/article/itenabled-business-transformation-from-automation-to-business-scope-redefinition.

6. Moore's Law, attributed to Gordon Moore, in a simplified version is that: "The processor speeds or the overall processing power for computers will double every two years." For an overview of the past fifty years of Moore's Law, see "50 Years of Moore's Law: Fueling Innovation We Love and Depend On," Intel, accessed July 18, 2016, www.intel.com/content/www/us/en/silicon-innovations/moores-law-technology.html. Although there's widespread acceptance within the technical community that this growth may come to an end in the near future, most believe that faster-cheaper-smaller computers will unleash new functionality.

7. Metcalfe's Law, attributed to Bob Metcalfe, states that: "The value of a telecommunications network is proportional to the square of the number of connected users of the systems (n^2)." Bob Metcalfe, "Metcalfe's Law: A Network Becomes More Valuable As It Reaches More Users," Infoworld 17, no. 40 (1995): 53–54. The same general idea can be extended to other networks such as social networks, devices connected to the Internet of Things, and other domains.

8. Gilder's Law, attributed to George Gilder, states that: "The bandwidth grows at least three times faster than the computing power." G. Gilder, "Fiber Keeps Its Promise: Get Ready. Bandwidth Will Triple Each Year for the Next 25," Forbes, April 1997, www.forbes.com/asap/97/0407/090.htm. Bandwidth refers to the amount of data that can be transmitted in a fixed amount of time, so this law implies that communication power doubles every six months. Cloud computing, which relies on a network of remote servers hosted on the Internet

to store, manage, and process data, is possible because fixed and mobile broadband networks are growing while their associated costs are decreasing.

9. Ed Zander, quoted in Ted C. Fishman, "What Happened to Motorola: How a Culture Shift Nearly Doomed an Iconic Local Company That Once Dominated the Telecom Industry," *Politics & City Life, Chicago Magazine*, accessed July 18, 2016, www.chicagomag.com/ Chicago-Magazine/September-2014/What-Happened-to-Motorola.

10. "Cracking the Digital Code: McKinsey Global Survey Results," McKinsey & Company, September 2015, accessed July 18, 2016, www. mckinsey.com/business-functions/business-technology/our-insights/ cracking-the-digital-code.

11. William Edwards Deming, *The New Economics: For Industry, Government, Education* (Cambridge, MA: MIT Press, 2000), 35.

12. William Bruce Cameron, *Informal Sociology: A Casual Introduction to Sociological Thinking* (New York: Random House, 1963), 13. Some people attribute this quote to Albert Einstein, but that has never been confirmed.

CHAPTER 1

1. "Model S Has You Covered," *Tesla Motors* (blog), March 19, 2015, www.teslamotors.com/blog/model-s-has-you-covered.

2. Danny Sullivan, "Google: 100 Billion Searches Per Month, Search to Integrate Gmail, Launching Enhanced Search App for iOS" *Search Engine Land* (blog), August 8, 2012, searchengineland.com/google-search-press-129925. An updated figure for 2014 puts this number at more than 2 trillion. Statistic Brain Research Institute, "Google Annual Search Statistics," accessed July 18, 2016, www.statisticbrain.com/google-searches.

3. Google reported the 1.4 billion number on September 12, 2015, during a press event, www.androidcentral.com/ google-says-there-are-now-14-billion-active-android-devices-worldwide

4. "Statistics," YouTube, accessed July 10, 2016, www.youtube.com/yt/press/ statistics.html.

5. Reported in *Fortune*, December 30, 2015, fortune.com/2015/12/30/ uber-completes-1-billion-rides.

6. Fitz Tepper, "Uber Has Completed 2 Billion Rides," *Tech Crunch*, July 18, 2016, techcrunch.com/2016/07/18/uber-has-completed-2-billion-rides.

7. One estimate puts the value of a taxi medallion in New York City at $1.3 million in 2013 and at $650,000 in 2015, representing a 50 percent loss of value. See Simon Van Zuylen-Wood, "The Struggles of New York City's

Taxi King," *Bloomberg BusinessWeek*, August 27, 2015, www.bloomberg.com/ features/2015-taxi-medallion-king.

8. Marriott Corporation, 2015 Annual Report.

9. "About Us," Airbnb, accessed August 26, 2016, www.airbnb.com/about/ about-us.

10. "About Us," Walmart, accessed June 29, 2016, corporate.walmart.com/ our-story.

11. Reported by Brian Olsavsky, CFO of Amazon. "Amazon.com (AMZN) Management on Q4 2015 Results—Earnings Call Transcript," Seeking Alpha, January 28, 2016, accessed July 27, 2016, seekingalpha.com/ article/3846086-amazon-com-amzn-management-q4-2015-results-earnings-call-transcript.

12. Alfred D. Chandler, *Scale and Scope: The Dynamics of Industrial Capitalism* (Cambridge, MA: Belknap Press, 1990).

13. See Over How Many Billions Served?, accessed July 27, 2016, overhow-manybillionserved.blogspot.com.

14. "Mark Zuckerberg, Moving Fast and Breaking Things," *Business Insider*, October 14, 2010, www.businessinsider.com/mark-zuckerberg-2010-10. In 2014, at the Facebook f8 conference, Zuckerberg amended his motto to "Move fast with stable infrastructure," which highlights the importance of maintaining stability at speed.

15. "Mark Zuckerberg's Letter to Investors: 'The Hacker Way,'" *Wired*, February 1, 2012, accessed July 26, 2016, www.wired.com/2012/02/ zuck-letter.

16. "Model S Has You Covered," *Tesla Motors* (blog), March 19, 2015, www.teslamotors.com/blog/model-s-has-you-covered.

17. Thomas Hout and George Stalk, "Time-Based Results," *bcg.perspectives* (blog), Boston Consulting Group, accessed July 18, 2016, www.bcgper-spectives.com/content/Classics/strategy_time_based_results/.

18. "Predix," GE Digital, accessed June 29, 2016, www.predix.com.

19. Pivot is a core concept of lean startups popularized by Steve Blank, "Why the Lean Start-Up Changes Everything," *Harvard Business Review*, May 2013, hbr.org/2013/05/why-the-lean-start-up-changes-everything. For more discussions about digital-era entrepreneurship, see steveblank.com.

20. For an overview, see "It's All A/Bout Testing: The Netflix Experimentation Platform," *Netflix Tech Blog*, April 29, 2016, techblog.netflix.com/2016/04/ its-all-about-testing-netflix.html.

21. Eric Ries lists five principles of a lean startup, and established companies have increasingly adopted them to accelerate their own innovation

process. For example, GE has developed its own version called FastWorks. See the Lean Startup website, theleanstartup.com.

22. Ray Kurzweil, "The Law of Accelerating Returns," Essays, Kurzweil Accelerating Intelligence, March 7, 2001, www.kurzweilai.net/the-law-of-accelerating-returns. All of the essays are worth reading.

23. There are several different forecasts. In a 2011 report, Cisco believed there would be 50 billion devices. See Dale Evans, *The Internet of Things: How the Next Evolution of the Internet Is Changing Everything*, White Paper produced for Cisco Internet Business Solutions Group (IBSG), April 2011 (San Jose, CA: Cisco, 2011), www.cisco.com/c/dam/en_us/about/ac79/docs/innov/IoT_IBSG_0411FINAL.pdf. Ericsson's target is around 26 billion, as stated in its 2015 *Ericsson Mobility Report: On the Pulse of the Networked Society* (Stockholm: Ericsson, 2015), www.ericsson.com/res/docs/2015/ericsson-mobility-report-june-2015.pdf.

CHAPTER 2

1. Gartner—an information technology research and advisory firm—has developed a graphical representation of the maturity and adoption of technologies and applications and how they are potentially relevant to solving real business problems and exploiting new opportunities. For more details, see "Gartner Hype Cycle," Research Methodologies, on the Gartner website, accessed June 30, 2016, www.gartner.com/technology/research/methodologies/hype-cycle.jsp.

2. For an overview of how some of the digital giants influence the network economy, see medium.com/inside-gafanomics. The website Gafanomics also provides useful reference points for four digital giants—Google, Apple, Facebook, and Amazon.

3. As a reference point, on July 13, 2016, Tesla was capitalized at $32.77 billion, whereas GM was at $47.16 billion and Ford Motors was at $53.55 billion. Computed from market data provided by Yahoo.

4. "The Automatic App + Adapter," Automatic Labs, accessed July 19, 2016, www.automatic.com/home.

5. "Safety & Efficiency: Truck Platooning," Peloton, accessed July 19, 2016, peloton-tech.com.

6. Mike Isaac and Michael J. de la Merced, "Uber Turns to Saudi Arabia for $3.5 Billion Cash Infusion," *New York Times*, June 1, 2016, www.nytimes.com/2016/06/02/technology/uber-investment-saudi-arabia.html. At the time, they reported Uber's valuation at $62.5 billion.

7. "GM and Lyft to Shape the Future of Mobility," General Motors, January 4, 2016, accessed July 19, 2016, www.gm.com/mol/m-2016-Jan-0104-lyft.html. GM invested $500 million in Lyft.

8. Jack Nicas and Jeff Bennett, "Alphabet, Fiat Chrysler in Self-Driving Cars Deal," *The Wall Street Journal*, May 3, 2016, www.wsj.com/articles/alphabet-fiat-chrysler-in-self-driving-cars-deal-1462306625.

9. "Toyota and Uber to Explore Ridesharing Collaboration," Toyota USA Newsroom, May 24, 2016, accessed July 19, 2016, corporatenews.pressroom.toyota.com/releases/toyota-uber-ridesharing-collaboration-may-24.htm.

10. Ingrid Lunden, "VW Invests $300M in Uber Rival Gett in New Ride-sharing Partnership," *Tech Crunch*, May 24, 2016, accessed July 19, 2016, techcrunch.com/2016/05/24/vw-invests-300m-in-uber-rival-gett-in-new-ride-sharing-partnership.

11. Platforms and platform-based business models have received a lot of attention over the last decade. For an overview of recent discussions in this area, see Geoffrey G. Parker, Marshall W. Van Alstyne, and Sangeet Paul Choudary, *Platform Revolution: How Networked Markets are Transforming the Economy and How to Make them Work for You* (New York & London: W.W. Norton & Co., 2016).

12. "Apple, Motorola & Cingular Launch World's First Mobile Phone with iTunes," Apple, September 7, 2005, www.apple.com/pr/library/2005/09/07Apple-Motorola-Cingular-Launch-Worlds-First-Mobile-Phone-with-iTunes.html.

13. "Introducing Vault0s: New Banking Model," ThoughtMachine, accessed July 14, 2016, www.beingtechnologies.com/#vault.

14. On June 28, 2016—as an illustrative reference point in time for comparison—Walmart was capitalized at $222 billion, whereas Amazon's value was at $334 billion—slightly more than 50 percent. My analysis is based on Yahoo Finance market data.

15. On June 28, 2016, Alibaba was capitalized at $190 billion. Yahoo Finance market data.

16. Frederick W. Taylor, *The Principles of Scientific Management* (New York: Harper & Brothers, 1911). This monograph is considered the foundation text for modern approaches to management and organizations.

17. See Michael E. Porter and James E. Heppelmann, "How Smart, Connected Products are Transforming Competition," *Harvard Business Review*, November 2014, accessed July 19, 2016, hbr.org/2014/11/how-smart-connected-products-are-transforming-competition. See also Bala Iyer and N.

Venkat Venkatraman, "What Comes After Smart Products" *Harvard Business Review*, July 1, 2015, hbr.org/2015/07/what-comes-after-smart-products.

18. "GE's Jeff Immelt on Digitizing in the Industrial Space," *McKinsey & Company*, October 2015, accessed July 19, 2016, www.mckinsey.com/business-functions/organization/our-insights/ges-jeff-immelt-on-digitizing-in-the-industrial-space.

19. See Robert Tercek, *Vaporized: Solid Strategies for Success in a Dematerialized World* (Vancouver: LifeTree Media, 2015) for an excellent treatment of the challenges faced by information-rich industries such as media and entertainment.

CHAPTER 3

1. Kyle Reissner, "Facebook for Industry—Mobility + Data + Analytics is a NOW Thing!," GE (blog), June 11, 2013, www.geautomation.com/blog/facebook-for-industry-mobility-data-analytics-is-a-now-thing.

2. Robin Wauters, "Vehicle Rental Giant Avis Acquires Car Sharing Company Zipcar for $500 Million," *Insider* (blog), Next Web, accessed July 19, 2016, thenextweb.com/insider/2013/01/02/vehicle-rental-giant-avis-acquires-car-sharing-company-zipcar-for-500-million/#gref. Avis announced its intent to acquire Zipcar in January 2013 and closed the deal in March 2013.

3. Robert Tercek provides an exhaustive list of "Uber for X" in different settings. See Tercek, *Vaporized*, 183–84.

4. For a good overview of Netflix's approach to recommendations, see Xavier Amatriain and Justin Basilico, "Netflix Recommendations: Beyond the 5 Stars (Part 1)," *Netflix Tech Blog*, April 6, 2012, techblog.netflix.com/2012/04/netflix-recommendations-beyond-5-stars.html.

5. Those interested to delve deeper into the reasons that Netflix chose Amazon Cloud, see "Four Reasons We Choose Amazon's Cloud as Our Computing Platform," *Netflix Tech Blog*, December 14, 2010, techblog.netflix.com/2010/12/four-reasons-we-choose-amazons-cloud-as.html. It took Netflix seven years to migrate to the cloud because the architects simply did not port the legacy problems to the cloud. See Yury Izrailevsky, "Completing the Netflix Cloud Migration," *Netflix Media Center* (blog), February 11, 2016, media.netflix.com/en/company-blog/completing-the-netflix-cloud-migration.

6. Netflix bid more than $100 million for *House of Cards* without requiring pilot episodes to be created and shown to test the audience. See Roberto Baldwin, "Netflix Gambles on Big Data to Become the HBO of Streaming," *Wired*, November 11, 2012, www.wired.com/2012/11/netflix-data-gamble/. This approach shows the possibility of mitigating risks more effectively through data analytics.

7. Experts genuinely had different views on the valuation of Uber in the beginning (and even now). See, for instance, Professor Aswath Damodaran's blogpost on October 12, 2015: aswathdamodaran.blogspot.com/2015/10/on-uber-rollercoaster-narrative-tweaks.html.

8. See Mark McClusky, "The Nike Experiment: How the Shoe Giant Unleashed the Power of Personal Metrics," *Wired*, June 22, 2009, www.wired.com/2009/06/lbnp-nike/.

9. Ibid.

10. Transcript from 2015 Nike, Inc. Investor Meeting, accessed July 27, 2016, s1.q4cdn.com/806093406/files/events/Documents/Mark-Parker_Final-Transcript-with-images.pdf.

11. Nike's chief operating officer, Erik Sprunk, remarked about the possible future of Nike as owning the digital intellectual property while exploring partnerships for the actual customized production of the shoe itself. Riley Jones, "A Nike Exec Says At-Home 3D-Printed Sneakers Aren't Far Away," *Complex*, October 5, 2015, www.complex.com/sneakers/2015/10/nike-eric-sprunk-3d-printed-sneakers.

12. Three acquisitions for about $710 million when Under Armour's revenues are around $4 billion is remarkable. See Adam Lashinsky, "Can Under Armour Score with Wearable Fitness Tech?," *Fortune*, January 14, 2016, fortune.com/2016/01/14/under-armour-fitness-wearables/.

13. Parmy Olson, "Silicon Valley's Latest Threat: Under Armour," *Forbes*, September 30, 2015, www.forbes.com/sites/parmyolson/2015/09/30/kevin-plank-under-armour-apps-technology/#5fda25004b25

14. Ibid.

15. "Under Armour and IBM to Transform Personal Health and Fitness, Powered by IBM Watson," news release on IBM website, January 6, 2016, www-03.ibm.com/press/us/en/pressrelease/48764.wss.

16. For a list of experiments announced in 2015, see Ford Smart Mobility newsletter, 2015, media.ford.com/content/dam/fordmedia/North%20America/us/2015/01/07/FordSmartMobilityMap.pdf.

17. "Mobility: The First Step" in *Sustainability Report 2014/15*, Ford Motor Company, accessed July 19, 2016, corporate.ford.com/microsites/sustainability-report-2014-15/mobility.html.

18. Ibid.

19. For Ford's announcement of a separate subsidiary to potentially expand its business scope beyond automobiles, see "Ford Smart Mobility LLC Established to Develop, Invest in Mobility Services; Jim Hackett Named Subsidiary Chairman," news release on Ford Motor Company Media

Center website, March 11, 2016, media.ford.com/content/fordmedia/fna/
us/en/news/2016/03/11/ford-smart-mobility-llc-established--jim-hackett-
named-chairman.html.

20. Bill Ford quoted in Joseph White, "Ford Chairman: 'We Are in an
Experimental Stage,'" *Forbes*, January 11, 2016, fortune.com/2016/01/11/
bill-ford-experiment-future-business-strategy.

21. "CEO Letter," GE 2014 *Annual Report*, GE, accessed July 19, 2016,
www.ge.com/ar2014/ceo-letter.

22. Ibid.

23. "GE's Jeff Immelt on "Digitizing in the Industrial Space," *McKinsey &
Company*, October 2015.

24. Google Health was shuttered in 2011, when Larry Page took
over as the CEO. See Brian Dolan, "10 Reasons Why Google
Health Failed," *MobiHealthNews*, June 27, 2011, mobihealthnews.
com/11480/10-reasons-why-google-health-failed.

25. See Verily website, accessed June 30, 2016, verily.com.

26. "Johnson & Johnson Announces Formation of Verb Surgical Inc., in
Collaboration with Verily," corporate announcement on Johnson
& Johnson website, December 10, 2015, www.jnj.com/news/all/
Johnson-johnson-Announces-Formation-of-Verb-Surgical-Inc-in-
Collaboration-With-Verily.

27. Alphabet's focus on aging is reflected in the setting up of Calico Labs.
See www.calicolabs.com.

28. Apple has taken a long view when it comes to health
care. See Arielle Duhaime-Ross, "Apple Dives Straight into
Health Care with Release of First CareKit Apps," April
28, 2016, www.theverge.com/2016/4/28/11510590/
carekit-apple-apps-glow-start-one-drop-diabetes-depression-pregrancy.

29. See HealthVault website, www.healthvault.com.

30. One of the key areas of emphasis for IBM Watson is health care. See
the IBM Watson Health website, www.ibm.com/smarterplanet/us/en/
ibmwatson/health.

31. Chris Messina of Uber has been credited with coining this phrase. See his
blog post of January 19, 2016, medium.com/chris-messina/2016-will-be-
the-year-of-conversational-commerce-1586e85e3991#.w496byozd.

32. Yongdong Wang, "Your Next New Best Friend Might Be a Robot,"
Nautilus (blog), February 4, 2016, nautil.us/issue/33/attraction/
your-next-new-best-friend-might-be-a-robot.

33. Ibid.

34. Steve Dent, "Ask Alexa to Add New Features to Your Amazon Echo," *engadget* (blog), June 28, 2016, www.engadget.com/2016/06/28/ask-alexa-to-add-new-features-to-your-amazon-echo.

35. Ibid.

36. "Swisscom Start-Up Challenge—A Unique Opportunity for Swiss Start-Ups," Swisscom, accessed July 19, 2016, www.swisscom.ch/en/business/start-up/swisscom-startup-challenge.html.

37. Ingrid Lunden, "BBVA Shuts In-House Venture Arm, Pours $250M into New Fintech VC Propel Venture Partners," *Tech Crunch*, February 11, 2016, accessed July 19, 2016, techcrunch.com/2016/02/11/bbva-shuts-in-house-venture-arm-pours-250m-into-new-fintech-vc-propel-venture-partners.

38. Dsamaniego, "Step Inside the Eureka Innovation Lab," *Unzipped* (blog), Levi Strauss & Co., September 22, 2014, accessed July 19, 2016, www.levistrauss.com/unzipped-blog/2014/09/step-inside-the-eureka-innovation-lab.

39. See Project Jacquard website, accessed July 19, 2016, atap.google.com/jacquard.

40. Warren Berger, "How Brainstorming Questions, Not Ideas, Sparks Creativity," Fast Company & Inc., June 6, 2016, www.fastcodesign.com/3060573/how-brainstorming-questions-not-ideas-sparks-creativity.

41. "Next47: Siemens Founds Separate Unit for Startups," Siemens Global, June 28, 2016, www.siemens.com/press/en/feature/2016/corporate/2016-06-next47.php?content[]=Corp.

42. By looking at information technology investments as strategic options, my co-author and I came up with the Strategic Options Navigator. Although it was developed during the craziness of the dot-com boom and the uncertainties associated with that period, the general idea of the various options still applies. See N. Kulatilaka and N. Venkatraman, "Strategic Options in the Digital Era," *Business Strategy Review* vol. 12, no. 4 (2001): 7–15. For a broader discussion on real options, see Martha Amram and Nalin Kulatilaka, *Real Options: Managing Strategic Investments in an Uncertain World* (Boston: HBS Press, 1998).

43. "Inside P&G's digital revolution," *McKinsey Quarterly*, November 2011, www.mckinsey.com/industries/consumer-packaged-goods/our-insights/inside-p-and-ampgs-digital-revolution.

CHAPTER 4

1. Adam Bryant, "Satya Nadella, Chief of Microsoft, on His New Role," *New York Times*, February 20, 2014, www.nytimes.com/2014/02/21/business/satya-nadella-chief-of-microsoft-on-his-new-role.html.

2. "Nest Labs Introduces World's First Learning Thermostat,"
   Nest, October 25, 2011, retrieved June 30, 2016, nest.com/press/
   nest-labs-introduces-worlds-first-learning-thermostat.

3. Ibid.

4. "Nest Protect: The Smoke + Carbon Monoxide Alarm Reinvented,"
   Nest, October 8, 2013, retrieved June 30, 2016, nest.com/ie/press/
   nest-protect-the-smoke-carbon-monoxide-alarm-reinvented.

5. Google acquired Nest Labs in January 2014. See "Welcome Home," Nest,
   January 13, 2014, retrieved June 30, 2016, nest.com/blog/2014/01/13/
   welcome-home.

6. See "Works with Nest," Nest, accessed June 30, 2016, nest.com/
   works-with-nest.

7. See "A Home that Adapts to You," Honeywell, accessed June 30, 2016,
   yourhome.honeywell.com/lyric.

8. "Leaving the Nest," Nest, June 3, 2016, accessed June 30, 2016, nest.com/
   blog/2016/06/03/leaving-the-nest.

9. "Accor Launches Its Digital Transformation 'Leading Digital Hospitality,'"
   Individual Shareholders' Webzine, Accor hotels, December 2014, accessed
   July 19, 2016, webzine-actionnaires.accorhotels-group.com/en/article/35/
   ACCOR_LAUNCHES_ITS_DIGITAL_TRANSFORMATION_LEADING_
   DIGITAL_HOSPITALITY.html.

10. Excerpted from "Mercedes-Benz at the 2012 International CES in Las Vegas:
    Mercedes-Benz Writes a New Chapter of the Declaration of Independence,"
    speech delivered at the 2012 International Consumer Electronics Show
    (CES), Las Vegas, January 11, 2012, accessed June 30, 2016. media.daimler.com/
    marsMediaSite/en/instance/ko/Mercedes-Benz-at-the-2012-International-
    CES-in-Las-Vegas-Mer.xhtml?oid=9917742.

11. Quoted in "BMWs That Can Buy Morning Coffee Tempt Apple, Google to
    Road," *Advertising Age*, July 6, 2015, accessed July 19, 2016, adage.com/article/
    digital/bmws-buy-morning-coffee-tempt-apple-google-road/299344.

12. See Google Self-Driving Car Project website, accessed July 27, 2016, www.
    google.com/selfdrivingcar. Specifically, it was reported, "We've self-driven
    over 1.5 million miles and have accumulated the equivalent of over 75 years
    of driving experience on the road (based on a typical American adult driving
    about 13,000 miles per year)."

13. Google was granted US patent number 8,078,349, "Transitioning a mixed-
    mode vehicle to autonomous mode" on December 13, 2011.

14. Marco della Cava and Brent Snavely, "Google, Fiat Ink Deal to Make
    100 Self-Driving Minivans," *Detroit Free Press*, May 3, 2016, accessed

July 26, 2016, www.freep.com/story/money/cars/2016/05/03/
google-fiat-chrysler-fca-self-driving-cars-chrysler-pacifica/83891878.

15. Lulu Yilun Chen, "Apple's Cook Struck $1 Billion Deal With
China's Didi in 22 Days," *Bloomberg*, May 13, 2016, accessed
July 26, 2016, www.bloomberg.com/news/articles/2016-05-13/
apple-s-cook-struck-1-billion-deal-with-china-s-didi-in-22-days.

16. Quoted in William Boston and Eric Sylvers, "Auto Makers Gear up to
Take on the Challenge from Google and Apple," *Wall Street Journal*, March
5, 2015, www.wsj.com/articles/auto-makers-gear-up-to-take-on-the-chal-
lenge-from-google-and-apple-1425417019.

17. See Sam Oliver, "Automakers in Geneva Cautious, 'Concerned' about
'Disruptive Interloper' Apple," Appleinsider, March 3, 2015, accessed
August 27, 2016, appleinsider.com/articles/15/03/03/automakers-in-gene-
va-cautious-concerned-about-disruptive-interloper-apple.

18. "BMWs That Can Buy Morning Coffee," *Advertising Age*.

19. "Connected Vehicles: Automobiles as Technology Platform with Don
Butler, Ford," interview by Michael Krigsman and Vala Afshar, CXO *Talk*,
podcast audio #109, May 8, 2015, www.cxotalk.com/connected-vehi-
cles-automobiles-technology-platform-don-butler-ford. Also discussions
with Ford Silicon Valley executives in 2013.

20. With so many different strategy pundits discussing the cycle of innova-
tion-disruption-transformation over the past two decades, there is no
excuse today for managers who do not understand and appreciate the
likely threats from disruption.

21. See W. Chan Kim and Renée Mauborgne, "What Is Blue Ocean
Strategy?," Blue Ocean Strategy, accessed July 20, 2016, www.blueoceans-
trategy.com/what-is-blue-ocean-strategy.

22. Geoffrey Moore introduces his hierarchy of powers involving five
areas—category power, company power, market power, offer power,
and execution power—that mutually reinforce each other to create a
distinct position among crowded options. See Geoffrey A. Moore, *Escape
Velocity: Free Your Company's Future from the Pull of the Past* (New York: Harper
Business, 2011).

23. "Nokia Completes Next Stage of Transformation with Agreement to Sell
HERE to Automotive Industry Consortium at an Enterprise Value of EUR
2.8 Billion," Nokia, August 3, 2015, accessed June 30, 2016, company.nokia.
com/en/news/press-releases/2015/08/03/nokia-completes-next-stage-
of-transformation-with-agreement-to-sell-here-to-automotive-industry-
consortium-at-an-enterprise-value-of-eur-28-billion.

24. Stephen Pulvirent, "TAG Heuer, Google Release First Swiss Luxury Smartwatch: All the Details," *Bloomberg Pursuits*, November 9, 2015, www. bloomberg.com/news/articles/2015-11-09/tag-heuer-connected-watch-from-google-and-intel-specs. See also "TAG Heuer Connected Watch with Intel Inside," Intel, accessed July 20, 2016, www.intel.com/content/www/us/en/wearables/tag-heuer-connected-watch.html.

25. For a list of acquisitions supporting Walmart's @WalmartLabs initiative, see "Acquisitions," @WalmartLabs, accessed June 30, 2016, www.walmartlabs. com/about/acquisitions.

26. "GM to Acquire Cruise Automation to Accelerate Autonomous Vehicle Development," announcement on GM website, March 11, 2016, accessed June 30, 2016, media.gm.com/media/us/en/gm/home.detail.html/content/Pages/news/us/en/2016/mar/0311-cruise.html.

27. As reported by Mike Ramsey, "Toyota Hires Entire Staff of Autonomous-Vehicle Firm," *Wall Street Journal*, March 9, 2016, www.wsj.com/articles/toyota-grabs-tech-talent-by-hiring-entire-jaybridge-staff-1457553870.

28. Bruce Upbin, "Monsanto Buys Climate Corp For $930 Million," *Forbes*, October 2, 2013, accessed July 27, 2016, www.forbes.com/sites/bruceupbin/2013/10/02/monsanto-buys-climate-corp-for-930-million/#6f28307a5ae1.

29. NYT *Innovation Report* 2014, accessed June 30, 2016, www.scribd.com/doc/224332847/NYT-Innovation-Report-2014.

CHAPTER 5

1. Theodore Levitt, "Marketing Myopia," *Harvard Business Review* 38 (July–August 1960), 24–47.

2. "About Facebook," Facebook, accessed June 30, 2016, www.facebook.com/facebook/info.

3. "About Tesla," Tesla Motors, accessed June 30, 2016, www.teslamotors.com/about.

4. Ibid.

5. Ibid.

6. Marta Falconi, "At Novartis, the Pill Is Just Part of the Pitch," *Wall Street Journal*, January 1, 2013, retrieved June 30, 2016, www.wsj.com/articles/SB10001424127887323635504578213801060673708.

7. Industry 4.0 is broadly about how a set of technologies affects manufacturing and extended supply chains across a set of companies. In my research on IT-enabled business transformation, I called this business network redesign in the 1990s. Now, a variety of technologies such as the Internet of

Things, data and analytics, and cloud computing allow for a more distributed network of production than was possible in the late twentieth century. See Germany Trade and Invest, *Industrie 4.0: Smart Manufacturing for the Future* (Berlin: Germany Trade and Invest, 2014), www.gtai.de/GTAI/Content/ EN/Invest/_SharedDocs/Downloads/GTAI/Brochures/Industries/indus-trie4.0-smart-manufacturing-for-the-future-en.pdf.

8. This is representative of the kind of assessments on the potential of Industry 4.0. See Cornelius Baur and Dominik Wee, "Manufacturing's Next Act," McKinsey & Company, June 2015, accessed July 20, 2016, www.mckinsey. com/business-functions/operations/our-insights/manufacturings-next-act.

9. I want to highlight that mastery of systems integration may involve a knowledge of computer science, but mastery of solution integration requires integrating knowledge from both computer science and business; it truly requires the integration of broad domain knowledge with deep technical knowledge.

10. IBM's integration of the assets and capabilities of the Weather Company illustrates the domain knowledge that it is acquiring to become a solution integrator. See the Weather Company website, business.weather.com.

11. IBM has also made significant acquisitions to bring new capabilities to the company as part of its reinvention. See, for example, "IBM Watson Health Closes Acquisition of Truven Health Analytics," IBM, April 8, 2016, www-03. ibm.com/press/us/en/pressrelease/49474.wss.

12. IBM and Apple have put aside their historical differences and forged a major alliance to develop the next generation of enterprise apps that could run on Apple iPads created for business use. See "IBM MobileFirst for iOS," IBM, accessed July 20, 2016, www.ibm.com/mobilefirst/us/en/mobilefirst-for-ios/

13. David Kesmodel, "Monsanto to Buy Climate Corporation for 930 Million; Loss Widens," *Wall Street Journal*, October 2, 2013, accessed July 27, 2016, www.wsj.com/articles/SB10001424052702304176904579111042198455468.

14. David S. Evans and Richard Schmalensee, *Matchmakers: The New Economics of Multisided Platforms* (Cambridge, MA: Harvard Business Review Press, 2016), 210.

15. GM's Maven is an initiative to provide seamless mobility, now with GM cars and associated technologies such as OnStar and links to Apple CarPlay and cellular connectivity. See "About," Maven, accessed June 30, 2016, www. mavendrive.com/#!.

16. Daimler's Moovel is a subsidiary that includes some of the company's acquisitions and seeks to reinvent urban mobility. See Moovel, accessed June 30, 2016, www.moovel.com/en/US.

17. Gautham Nagesh, "Mary Barra's Road Map for GM Centers on Customer Data, Connectivity," *Wall Street Journal*, October 25, 2015, www.wsj.com/articles/mary-barras-road-map-for-gm-centers-on-customer-data-connectivity-1445824801.

18. Ibid.

19. On January 6, 2015, Ford published a series about mobility experiments on its media center newsfeed. See, for example, "Mobility Experiment: Painless Parking," Ford Motor Company, January 6, 2015, media.ford.com/content/fordmedia/fna/us/en/news/2015/01/06/mobility-experiment-painless-parking-london.html.

20. Horst W.J. Rittel and Melvin M. Webber, "Dilemmas in a General Theory of Planning," *Policy Sciences* 4 (1973), 155–69.

21. John C. Camillus, *Wicked Strategies* (Toronto: University of Toronto Press, 2016).

22. The Institute of Design at Stanford (d.school) is at the forefront of innovation in the interdisciplinary area of design thinking. See dschool.stanford.edu.

23. IDEO is the reference company for design thinking. See www.ideo.com.

24. See "Our Story," Embrace, accessed July 21, 2016, embraceglobal.org/about-us.

25. Ranjay Gulati, *Reorganize for Resilience: Putting Customers at the Center of Your Business* (Cambridge, MA: Harvard Business Review Press, 2010).

26. One of the most famous examples involved solving the problem of finding longitude while at sea. Based on conventional thinking, Isaac Newton and others expected the solution to come from astronomy, but the final solution was at the intersection of chronometry and astronomy. John Harrison, a carpenter and clockmaker, has been credited with contributing to the solution that we still use today. See "The History," Longitude Prize, accessed July 21, 2016, longitudeprize.org/history. See also Dava Sobel, *Longitude: The True Story of a Lone Genius Who Solved the Greatest Scientific Problem of His Time* (New York: Walker Books, 2007).

27. P&G has been working with the connect + develop program since 2000. See www.pgconnectdevelop.com.

28. See "Innovate with InnoCentive," InnoCentive, accessed July 21, 2016, www.innocentive.com.

29. Quoted in Steve Lohr, *Data-Ism: Inside the Big Data Revolution* (New York: Harper Business, 2015), 61.

CHAPTER 6

1. Quoted in Carmine Gallo, "Steve Jobs: The World's Greatest Storyteller," *Forbes*, October 8, 2015, www.forbes.com/sites/carminegallo/2015/10/08/steve-jobs-the-worlds-greatest-business-storyteller/#325cd8d6bfe3.

2. For an academic treatment on the shift in industry architecture, see Timothy F. Bresnahan and Shane Greenstein, "Technological Competition and the Structure of the Computer Industry," *Journal of Industrial Economics*, vol. 47 no. 1 (1999), 1–40. See also Andy Grove's discussion of the shift from vertical to horizontal integration of the computer industry in his book *Only the Paranoid Survive: How to Exploit the Crisis Points that Challenge Every Company* (New York: Doubleday, 1999), 39–51.

3. Bill Gates, in his letter dated June 25, 1985, implored Apple to establish the Macintosh ecosystem. "As the independent investment in a 'standard' architecture grows, so does the momentum for that architecture. The industry has reached the point where it is now impossible for Apple to create a standard out of their innovative technology without support from, and the resulting credibility of other personal computer manufacturers. Thus, Apple must open the Macintosh architecture to have the independent support required to gain momentum and establish a standard… Microsoft is very willing to help Apple implement this strategy. We are familiar with the key manufacturers, their strategies and strengths. We also have a great deal of experience in OEMing system software." See scripting.com/specials/gatesLetter/text.html.

4. See Merchant Customer Exchange (MCX), accessed on June 30, 2016, www.mcx.com.

5. "Walmart Introduces Walmart Pay," *Business Wire*, December 10, 2015, www.businesswire.com/news/home/20151209006572/en/Walmart-Introduces-Walmart-Pay.

6. Steve Cheney, "System Wide Network Effects in Mobile," blog post on his website, July 6, 2015, stevecheney.com/system-wide-network-effects-in-mobile.

7. Ibid.

8. In the 1980s, political scientist Joseph S. Nye Jr. coined the term "soft power," the economic equivalent of market power as applied in foreign affairs. I find this concept an apt description for the orchestration of ecosystems. Joseph S. Nye Jr., *Soft Power: The Means to Success in World Politics* (New York: Public Affairs, 2004).

9. An illustrative case is the tension between Apple and Spotify. See John Paczkowski, "Apple Slams Spotify, Says App Already Violates App

Store Rules," July 1, 2016, accessed July 12, 2016, www.buzzfeed.com/
johnpaczkowski/apple-fires-back-at-spotify-for-asking-for-preferential-
trea?utm_term=.ecpnmOl8J#.rl9z4OIxq.

10. See Chapter 5, note 4.

11. See "Capital Markets Day—2015 London, September 9th," presentations and discussions on ABB website, accessed July 25, 2016, new.
abb.com/investorrelations/financial-results-and-presentations/
capital-markets-day-2015.

CHAPTER 7

1. Lawrence M. Fisher, "Preaching Love Thy Competitor," *New York Times*,
March 29, 1992, accessed July 22, 2016, www.nytimes.com/1992/03/29/
business/preaching-love-thy-competitor.html. See also Charles Bruno,
"Big Red Keeps Rolling," *Network World*, October 4, 1995, 50.

2. Quoted in Walter Isaacson, *Steve Jobs* (New York: Simon & Schuster, 2011).

3. Steve Jobs comments during Apple's all-hands meeting in January 2010 as
reported in the *New York Times* on March 14, 2010. Brad Stone and Miguel
Helft, "Apple's Spat with Google Is Getting Personal," *New York Times*,
March 14, 2010, B1, www.nytimes.com/2010/03/14/technology/14brawl.
html?pagewanted=all&_r=0.

4. As quoted by Joel Rosenblatt and Adam Satariano, "Google Paid Apple
$1 Billion to Keep Search Bar on iPhone," *Bloomberg Technology* (blog),
January 21, 2016, accessed July 1, 2016, www.bloomberg.com/news/arti-
cles/2016-01-22/google-paid-apple-1-billion-to-keep-search-bar-on-iphone.
Based on court documents filed during a lawsuit between Oracle and
Google over Java software.

5. Quoted in Charles Bruno, "Big Red Keeps Rolling," *Network World*, October
4, 1995, 50.

6. Adam Brandenburger and Barry Nalebuff, *Co-opetition: A Revolutionary
Mindset that Combines Competition and Cooperation* (New York: Currency
Doubleday, 1996).

7. Quoted in Russell Hotten, "Carmakers Face Challenge from Google and
Apple," BBC News, March 4, 2015, accessed July 27, 2016, www.bbc.com/
news/business-31720645.

8. "TAG Heuer, Google and Intel Announce Swiss
Smartwatch Collaboration," Intel, March 19, 2015, accessed
July 22, 2016, newsroom.intel.com/news-releases/
tag-heuer-google-and-intel-announce-swiss-smartwatch-collaboration.

9. Joe Belfiore, "Lenovo introduces great new PCs to bring out the best in Windows 10 and connectivity," blog post on Windows website, October 19, 2015, accessed July 22, 2016, blogs.windows.com/windowsexperience/2015/10/19/lenovo-introduces-great-new-pcs-to-bring-out-the-best-in-windows-10-and-connectivity.

10. Tom Warren, "Lenovo Refused to Sell Microsoft's Surface Because They're 'Competitors,'" *Verge*, October 16, 2015, accessed July 22, 2016, theverge.com/2015/10/16/9549183/lenovo-refused-to-sell-microsoft-surface-pro-3.

11. Academic research has focused on capabilities during the last two decades. Prof. Paul Pavlou and Omar El Sawy introduce the ideas of improvisational capabilities in the context of product development, which could be usefully extended to multi-firm settings under coopetitive conditions. For their original work, see Paul A. Pavlou and Omar A. El Sawy, "The Third Hand: IT-Enabled Competitive Advantage in Turbulence through Improvisational Capabilities," *Information Systems Research*, vol. 21, no. 3 (2008): 443–71.

12. "Google: signs of mortality?" in WPP, *Annual Report and Accounts 2009*, accessed July 22, 2016, www.wpp.com/annualreports/2009/what-we-think/new-markets-new-media-and-consumer-insight-by-sir-martin-sorrell/google-signs-of-mortality.html. In this document, Sir Martin Sorrell, chief executive officer, uses the spelling "frienemy," which was modified by the *Financial Times* to be "frenemy."

13. Sir Martin Sorrell, "What We Think," in WPP, *Annual Report 2012*, p. 88, accessed July 22, 2016, www.wpp.com/wppataglance/2013/pdf/what-we-think/sirmartin-sorrells-article.pdf.

14. Sir Martin Sorrell, in WPP, *Annual Report and Accounts 2015*, accessed July 22, 2016, www.wpp.com/annualreports/2015/what-we-think/the-case-for-sticking-your-neck-out/#data-big-getting-bigger.

15. Babolat introduced Babolat Play to help tennis players track, analyze, and evaluate their game at a very detailed level. See "Babolat Play," Babolat, accessed July 1, 2016, en.babolatplay.com.

CHAPTER 8

1. IBM's Deep Blue beat the world chess champion in six games. Two wins for IBM, one for Gary Kasparov, and three draws. See "Deep Blue," IBM 100, accessed July 1, 2016, www-03.ibm.com/ibm/history/ibm100/us/en/icons/deepblue.

2. At the end of the game challenge, when IBM Watson won, Ken Jennings, the seventy-four-time winner of this popular TV quiz show, wrote: "I, for one,

welcome our new computer overlords." See Adam Gabbatt, "IBM Computer Watson Wins Jeopardy Clash," *Guardian*, February 17, 2011, theguardian.com/technology/2011/feb/17/ibm-computer-watson-wins-jeopardy.

3. IBM formed a separate unit, IBM Watson Group, focused on developing and commercializing cloud-delivered cognitive solutions and committed $100 million for venture investments to accelerate the formation of ecosystems. See "IBM Forms New Watson Group to Meet Growing Demand for Cognitive Innovations," IBM, January 9, 2014, accessed July 1, 2016, www-03.ibm.com/press/us/en/pressrelease/42867.wss.

4. Virginia M. Rometty, "Chairman's Letter," in 2015 *Annual Report*, IBM, accessed July 27, 2016, www.ibm.com/annualreport/2015/chairmans-letter.

5. Here I have used "IBM Watson" because it is today's leading example. However, I mean cognitive machines in general.

6. Satya Nadella, "The Partnership of the Future," *Slate*, June 28, 2016, accessed July 24, 2016, www.slate.com/articles/technology/future_tense/2016/06/microsoft_ceo_satya_nadella_humans_and_a_i_can_work_together_to_solve_society.html.

7. For more details on cloud robotics, see "Cloud Robotics and Automation," University of California, Berkeley, accessed July 27, 2016, project on cloud robotics and automation. goldberg.berkeley.edu/cloud-robotics.

8. Carl B. Frey and Michael A. Osborne, "The Future of Employment: How Susceptible are Jobs to Computerization?" Oxford Martin School working paper, September 17, 2013, www.oxfordmartin.ox.ac.uk/downloads/academic/The_Future_of_Employment.pdf.

9. "EPS Estimates Down for J.M. Smucker in Past Month," *Forbes*, October 12, 2015, accessed July 24, 2016, www.forbes.com/sites/narrativescience/2015/10/12/eps-estimates-down-for-j-m-smucker-in-past-month/#424e0f7a45da.

10. "What Is Watson?" IBM Watson, accessed July 24, 2016, www.ibm.com/watson/what-is-watson.html.

11. Tom Davenport and Julia Kirby, *Only Humans Need Apply: Winners & Losers in the Age of Smart Machines* (New York: Harper Business, 2016). The authors discuss in detail the needs and challenges of augmentation. And they discuss the current state of "smartness" of smart machines. The authors also discuss five ways for human talent to be continually employable: stepping up to develop big picture insights, stepping aside to areas that computers are not good at, stepping in as computer programmers, stepping narrowly into specialized areas, and stepping forward to develop new methods and systems.

12. Bureau of Labor Statistics, US Department of Labor, economic news release, July 12, 2016, www.bls.gov/news.release/jolts.nr0.htm.

13. Jerry Kaplan, *Humans Need Not Apply: A Guide to Wealth and Work in the Age of Artificial Intelligence* (New Haven, CT: Yale University Press, 2015), 5.

14. Ibid.

15. Rethink Robotics in Boston is among the leaders in next-generation collaborative robots. See www.rethinkrobotics.com.

16. Gary Kasparov, "The Chess Master and the Computer" *New York Review of Books*, February 11, 2010, www.nybooks.com/articles/2010/02/11/the-chess-master-and-the-computer.

17. Interview quoted in Sam Byford, "DeepMind Founder Demis Hassabis on How AI Will Shape the Future," *Verge*, March 10, 2016, www.theverge.com/2016/3/10/11192774/demis-hassabis-interview-alphago-google-deepmind-ai.

18. Mark Zuckerberg, post on Facebook page, January 26, 2016, www.facebook.com/zuck/posts/10102619979696481.

19. Airbnb acquired the talent of ChangeCoin in the form of "aqui-hire." See Ian Karr and Joon Ian Wong, "Airbnb Just Acquired a Team of Bitcoin and Blockchain Experts," *Quartz*, April 12, 2016, accessed July 24, 2016, qz.com/657246/airbnb-just-acquired-a-team-of-bitcoin-and-blockchain-experts.

20. Erik Brynjolfsson and Andrew McAfee, *The Second Machine Age: Work Progress and Prosperity in a Time of Brilliant Technologies* (New York: W.W. Norton, 2014).

21. A 2015 analysis by L2 reported that more than one thousand executives have left the top-branded consumer industry for the digital giants. Although it does not presage a larger trend in any one specific direction, this is an issue of relevance as talented humans capable of working with powerful machines in different ways. Sir Martin Sorrell cited in Mindi Chahal, "Sir Martin Sorrell on Digital, Cost-Cutting and Client-Agency Relationships," *Marketing Week*, February 4, 2016, accessed July 24, 2016, www.marketingweek.com/2016/02/04/sir-martin-sorrell-on-digital-cost-cutting-and-client-agency-relationships.

22. Laszlo Bock, *Work Rules! Insights from Inside Google That Will Transform How You Live and Lead* (New York: Hachette Group, 2015), 371.

23. N. Venkatraman, "Alphabet Isn't a Typical Conglomerate," *Harvard Business Review Online*, August 2015, accessed July 11, 2016, hbr.org/2015/08/alphabet-isnt-a-typical-conglomerate.

CHAPTER 9

1. Peter F. Drucker, "The Theory of the Business," *Harvard Business Review*, September–October 1994, 95–104.

2. James G. March, "Exploration and Exploitation in Organizational Learning," *Organization Science*, vol. 2, no. 1 (1991): 71–87. This classic academic work has stimulated a wide body of research on organizational adaptation.

3. D.A. Levinthal and J.G. March, "The Myopia of Learning," *Strategic Management Journal*, vol. 14 (1993), 95–112.

4. H.I. Ansoff, R.P. Declerck, and R.L. Hayes, eds., *From Strategic Planning to Strategic Management* (London: John Wiley & Sons, 1976). This was one of the first books that I read as a doctoral student in strategic management, and the ideas of weak signals have been deeply ingrained in my thinking ever since.

5. Douglas K. Smith and Robert C. Alexander, *Fumbling the Future: How Xerox Invented, Then Ignored the First Personal Computer* (New York: William Morrow & Co., 1988).

6. Louis M. Gerstner, *Who Says Elephants Can't Dance: Inside IBM's Historic Turnaround* (New York: Harper Collins, 2002).

7. Clayton M. Christensen, *The Innovator's Dilemma: When New Technologies Cause Great Firms to Fail* (Boston: Harvard Business Press, 1997).

8. Howard Yu, "What Pokémon Go's Success Means for the Future of Augmented Reality," *Fortune*, July 23, 2016, accessed July 25, 2016, fortune.com/2016/07/23/pokemon-go-augmented-reality.

9. Sunny Dhillon quoted in Igal Raichelgauz, "Pokémon Go is Nice, but Here's What *Real* Augmented Reality Will Look Like," *VentureBeat*, July 24, 2016, accessed July 25, 2016, venturebeat.com/2016/07/24/pokemon-go-is-nice-but-heres-what-real-augmented-reality-will-look-like.

10. Claude Bernard, *An Introduction to the Study of Experimental Medicine* (US: Henry Schuman, Inc., 1949), 18. Digitized by the Internet Archive in 2014, archive.org/details/b21270557.

11. Jeff Bezos, 2014 Letter to shareholders, accessed July 25, 2016, www.sec.gov/Archives/edgar/data/1018724/000119312514137753/d702518dex991.htm.

12. See Local Motors 3D (LM3D), accessed July 25, 2016, localmotors.com/3d-printed-car.

13. Manu S. Mannoor, Ziwen Jiang, Teena James, Yong Lin Kong, Karen A. Malatesta, Winston O. Soboyejo, Naveen Verma, David H. Gracias, and

Michael C. McAlpine, "3D Printed Bionic Ears," *Nano Letters*, vol. 13, no. 6 (2013): 2634–39.

14. "Aurora Flight Sciences and Stratasys Deliver World's First Jet-Powered, 3D Printed UAV in Record Time," Stratasys, November 9, 2015, accessed July 25, 2016, investors.stratasys.com/releasedetail.cfm?ReleaseID=941406.

15. Larry Page, "Larry's Alphabet Letter," on Alphabet Investor Relations page, accessed July 25, 2016, abc.xyz/investor/founders-letters/2015.

16. Niccolò Machiavelli, *The Prince*, English translation by Ninian H. Thomson, (New York: P.F. Collier & Son, 1909), originally published in Florence, 1513.

17. Gordon R. Sullivan and Michael V. Hopper, *Hope is Not a Method: What Business Leaders Can Learn from America's Army* (New York: Broadway Books, 1997).

18. Ibid.

19. John F. Kennedy, "266—Address in the Assembly Hall at the Paulskirche in Frankfurt," transcript of speech from June 25, 1963, American Presidency Project, accessed July 26, 2016, www.presidency.ucsb.edu/ws/?pid=9303.

20. Jesse Willms, "R3 Blockchain Development Initiative Grows to 22 Banks Worldwide," *Bitcoin Magazine*, September 29, 2015, accessed July 25, 2016, bitcoinmagazine.com/articles/r-blockchain-development-initiative-grows-to-banks-worldwide-1443553081.

CHAPTER 10

1. Jeff Bezos, 1997 Letter to shareholders, accessed August 2, 2016, www.sec.gov/Archives/edgar/data/1018724/000119312514137753/d702518dex991.htm.

2. David L. Rogers, *The Digital Transformation Playbook* (New York: Columbia Business School Publishing. 2016). Pay particular attention to his discussions of *convergent* and *divergent* experiments in Chapter 5: "For innovations intending to improve your existing core business, you are more likely to rely on convergent experiments. For innovations intending to develop new business areas and generate substantially new products, services, or processes, you are more likely to depend on divergent experiments."

3. George Westerman, Didier Bonenet, and Andrew McAfee. *Leading Digital: Turning Technology into Business Transformation* (Boston: Harvard Business School Press, 2014).

4. IBM Institute for Business Value, *Device Democracy: Saving the Future of the Internet of Things* (Somers, NY: IBM Corporation, 2015), available online at www-935.ibm.com/services/us/gbs/thoughtleadership/internetofthings.

5. "The Trust Machine," *Economist* (print edition), October 31, 2015, available online at www.economist.com/news/leaders/21677198-technology-behind-

bitcoin-could-transform-how-economy-works-trust-machine (pay wall or registration).

6. Peter Thiel, *Zero to One: Notes on Startups, or How to Build the Future* (New York: Crown Business, 2014), 151.

7. Julia La Roche, "Goldman Sachs Just Pulled a Silicon Valley Move," *Business Insider*, August 12, 2015, accessed July 25, 2016, www.businessinsider.com/goldman-sachs-is-making-its-trading-tech-open-source-2015-8. SIMON stands for Structured Investment Marketplace and Online and is helping the company to redefine its skill base: of the roughly 35,000 people on Goldman's payroll, 9,000 are now programmers or engineers.

8. Nathaniel Popper, "A Gay, Latino Partner Tests Goldman's Button-Down Culture," *New York Times*, April 1, 2016, accessed July 31, 2016, www.nytimes.com/2016/04/03/business/dealbook/goldmans-tech-chief-pushes-the-bank-to-be-more-open-like-him.html?_r=0.

9. Arun Sundararajan, *The Sharing Economy: The End of Employment and the Rise of Crowd-Based Capitalism* (Cambridge, MA: The MIT Press, 2016).

10. For example, my research on software ecosystems carried out with my colleagues Bala Iyer and Chi-Hyon Lee showed how a set of network metrics help you understand the dynamics of ecosystems in software. See Bala Iyer, Chi-Hyon Lee, and N. Venkatraman, "Managing in a 'Small World Ecosystem': Lessons from the Software Sector," *California Management Review*, vol. 48, no. 3 (2006): 28–47.

11. A.R. Guess, "How PayPal Is Using Deep Learning to Root Out Fraud," *Dataversity*, March 10, 2015, accessed July 26, 2016, www.dataversity.net/paypal-using-deep-learning-root-fraud.

12. Romain Dillet, "BMW, Mobileye and Intel Are Building a Full Self-Driving Car for 2021," *Tech Crunch*, July 1, 2016, accessed July 25, 2016, techcrunch.com/2016/07/01/bmw-mobileye-and-intel-are-building-a-full-self-driving-car-for-2021.

13. The question of whether computers will take over from humans has been the subject of much debate. Two ends of the spectrum are represented by Jerry Kaplan, *Humans Need Not Apply: A Guide to Wealth and Work in the Age of Artificial Intelligence* (New Haven, CA: Yale University Press, 2015) and Tom Davenport and Julia Kirby, *Only Humans Need Apply: Winners and Losers in the Age of Smart Machines* (New York: HarperBusiness, 2016). In my opinion, the answer lies somewhere in between, as humans are inherently more creative than machines.

14. Thiel, *Zero to One*, 151.

15. Andrew Grove, *Only the Paranoid Survive: How to Exploit the Crisis Points That Challenge Every Company* (New York: Doubleday Business, 1996).

16. Stephen Elop, as reported by Chris Ziegler, "Nokia CEO Stephen Elop Rallies Troops in Brutally Honest 'Burning Platform' Memo? (Update: It's Real!)," *engadget*, February 8, 2011, accessed July 27, 2016, www.engadget.com/2011/02/08/ nokia-ceo-stephen-elop-rallies-troops-in-brutally-honest-burnin.

17. Kevin J. Delaney, "No One Should Have The Word 'Strategy' in Their Job Title," *Quartz*, May 12, 2016, accessed July 25, 2016, qz.com/680887/ no-one-should-have-the-word-strategy-in-their-job-title.

INDEX

automation, 173, 174–75, 183
automotive industry: business model
  reinvention, 108, 110–13; co-creation,
  165; in collision at the core phase, 47–48,
  88–92; conceptualization of cars, 88–89;
  control of apps, 92; digital giants in, 158;
  ecosystem engagement strategies, 134–
  37; ecosystems in, 145; experimentation,
  89–90; Ford challenge-focused
  experimentation example, 66–67;
  human resources, 177; platforms in, 109;
  reinvention at the root in, 109; response
  to digitization, 91–92; and Rules Matrix,
  214; self-driving cars, 90, 225, 256nn12–
  13; types of players in, 36–39; Uber
  experimentation on the edge example,
  56–57; value-to-cost advantage, 94. *See
  also* Daimler; Ford Motors; General
  Motors; Tesla Motors; Uber

B

Babolat, 165, 263n15
Ballmer, Steve, 11
bandwidth, 247n8
banking industry, *see* financial services
  industry
Barra, Mary, 111, 112
batteries, 102–3, 145
BBVA, 76
Belfiore, Joe, 160
Benioff, Marc, 192
Berger, Warren, 77
Bernard, Claude, 201
Bezos, Jeff, 194, 201, 203, 207
Bitcoin, 205–6
Biver, Jean-Claude, 158–59
BlackBerry, 10, 41, 80, 93, 132, 157, 228
Blank, Steve, ? 49n19
blockchain, 141, 147, 184, 192, 215
BMW, 14, 36, 95, 97, 225
Bock, Laszlo, 186
Boeing, 202, 213
Bosch: as digital company, 15, 36–37; as
  digital leader, 46, 48; orchestration
  intentions, 136, 140; partnership with
  ABB, 146; rules for, 213; use of data and
  analytics, 105
boundaries, transcending, 119
Brandenburger, Adam: *Co-opetiton*, 151–52
Brynjolfsson, Erik, 185

Buffett, Warren, 141
business, theory of, 190
business models: approach to, 109–10,
  128; automotive industry example,
  110–13; choosing, 128–29, 131–32; life
  cycle of, 96; lightbulb example, 129–32;
  platform-focused, 108–9, 128, 130–31,
  251n11; product-focused, 107–8, 129; for
  reinvention, 98–99, 113; service-focused,
  108, 130; solution-focused, 109, 128, 131;
  types of, 107. *See also* organizational
  structure

C

Cameron, Bruce, 13
Camillus, John: *Wicked Strategies*, 116
capabilities: distinctive, 50–51, 163–64,
  187; improvisational, 263n11
capability co-creation, 153, 155, 162–63,
  216–17
car rental industry, 57, 252n2
challenge-focused experimentation, 61,
  66–69
Chandler, Alfred, 20
ChangeCoin, 184, 265n19
chatbots, 42, 71, 109
Chavez, Rick, 117
Christensen, Clayton, 199
cloud robotics, 174
co-creation zone, 159–61, 164–65, 216.
  See also capability co-creation
coexistence strategies, 93–96. *See also*
  coopetition; relationships
cognitive computing, *see* powerful machines
collision at the core: introduction, 42–44,
  80; acquisitions, 95, 97; alliances, 95;
  automotive industry example, 47–48,
  88–92; business model change, 98–99;
  coexistence response, 93–96; cost
  advantage response, 94; divestment
  and refocus on digital core, 96–97;
  ecosystem impacts, 86; failure to
  respond, 93; Honeywell example, 83–87;
  in hotel industry, 87–88; inevitability
  of, 80–81; morphing response, 96–99;
  organizational structure collisions,
  82–83, 87; process of, 113–14; rules for,
  223; signs of, 81–83; strategy collisions,
  81–82
Comcast, 48, 102, 194–95

with Amazon, 161; data collection, 21; ecosystem orchestration, 219–20; experimentation on the edge by, 58–61; *House of Cards*, 59–60, 176, 252n6; impact of, 48; migration to cloud, 252n5; use of scale-scope-speed nexus, 29

Netflix Prize, 59

Newton, Isaac, 260n26

New York Times, 98, 101

next47 (Siemens), 78, 229

Nike, 61–63, 253n11

Nokia, 10, 11, 26, 95, 132, 195, 228

non-linearity, 29–30

Noorda, Ray, 148, 150

Novartis, 15, 104–5, 145, 203

Nye, Joseph S., Jr., 261n8

O

objectives, clarifying, 229–30

observation, 73–77

Oncology Expert Advisor, 176

Only Humans Need Apply (Davenport and Kirby), 264n11

Only the Paranoid Survive (Grove), 231

orchestration, 126–27, 133–34, 136–37, 140–44

organizational structure, 8, 82–83, 87, 183–86. *See also* business models

P

Page, Larry, 203

Palantir Technologies, 35, 172, 222

Parker, Mark, 62, 63

participation, 126, 134, 136–37

passions, identifying, 118–19

Pavlou, Paul, 263n11

Paypal, 35, 222

Peloton Technology, 37

personal health, *see* health care

personalization, 59, 60, 65. *See also* recommendation engines

philanthropy, 141

pivot, 28–29, 249n19

Plank, Kevin, 63–64, 64–65

platforms: business model for, 108–9, 128, 130–31, 251n11; definition of, 40, 126; dominance of, 133, 144; value in, 108–9. *See also* ecosystems

PlayStation (Sony), 27

powerful machines: introduction, 167, 169; added value and capabilities from, 172–73; amplification, 173, 176–78, 180, 182–83, 227; artificial general intelligence (AGI), 178; augmentation, 173, 175–76, 183, 222–23; automation, 173, 174–75, 183; competition within field, 178; games played against, 167–68, 181, 182, 263nn1–2; human partnership with, 180, 181–82, 185–86, 226–27, 268n13; job replacement by, 170; job support from, 170–71; keeping pace with, 180–81; as mainstream, 168–69; *Only Humans Need Apply* (Davenport and Kirby) on, 264n11; organizational design for, 183–86, 187; rethinking work to account for, 169–72, 173–74, 183; robots, 179–80; as synthetic intellects, 178–79; taking advantage of, 171–72, 217–18

problem framing, 116–19

problem solving, 100–101, 103, 104, 119–20

Procter & Gamble (P&G), 19, 78–79, 119, 260n27

product-focused business model, 107–8, 129

Project Jacquard (Google), 76, 165

Q

Quill (Narrative Science), 175–76

R

recommendation engines, 58–59, 60, 64–65, 174. *See also* personalization

reinvention at the root: introduction, 44–46, 100; automotive industry example, 110–13; business models for, 107–10, 113; co-creation zone, 159–61; Facebook example, 101–2; health care example, 104–5; IBM example, 105–6; identifying passions, 118–19; Industrial Internet example, 105; key questions, 104; media and entertainment industries example, 48; Monsanto example, 106–7; need for repetition, 120; outside-in approach to solutions, 118; partnerships, 119–20; problem framing, 116–19; problem solving, 100–101, 103, 104, 119–20; process of, 114, 115–16; relevance during, 114–15; rules for, 227; shift in business logic, 115; Tesla example, 102–3; transcending boundaries, 119